COVID
OPERATION

COVID
OPERATION

What Happened, Why It Happened, and What's Next

PAMELA A. POPPER

SHANE D. PRIER

Paperback ISBN: 978-1-63337-443-0
E-book ISBN: 978-1-63337-444-7
LCCN: 2020919525

Printed in the United States of America

1 3 5 7 9 10 8 6 4 2

CONTENTS

INTRODUCTION

by Pam Popper

ON MARCH 10, 2020, I posted a video on my YouTube Channel titled "What You Need to Know About the Coronavirus." I was prompted to create the video because we had hosted a large event at our office on March 7, and a few people decided not to attend because they were afraid of "the virus." We had planned a large event with a guest speaker on April 2 and we were getting calls about whether this was going to take place because of "the virus." I was curious – flu season is an annual event, and I had never heard of anything being cancelled as a result of it.

I started my investigation with a lot of skepticism about the seriousness of "the virus" due to my investigation of the H1N1 (Swine) flu debacle in 2009-2010. Both the CDC and WHO were found to have engaged in unethical practices in order to turn run-of-the-mill flu into a pandemic. Drug companies benefitted financially, millions of people were vaccinated without cause, and without evidence that the vaccine was effective since it was not clinically tested. Could this type of deception be behind the growing concern about this virus?

It seemed that this was the case. At the time my video was posted, the number of cases and deaths from COVID-19 worldwide was a small fraction of what was considered normal for flu season. But an already disproportionate response had begun, and I reported some of the potential motivations for this exaggerated response in my video.

I have been posting videos every Tuesday and Thursday for many years, many of which have been considered "controversial." Nasty comments, both from consumers and health professionals, have been a consistent factor in my life for my entire 25 years in healthcare. But the response to this post was different in several ways. This time there were many more critics, and they were considerably more hostile. People accused me of putting the lives of others in danger by minimizing the risk of death from COVID-19. "You will ultimately be responsible for killing people," one person wrote. Some sent condescending emails stating that I was not qualified to talk about this topic. "You're in over your head on this one," one viewer wrote. "I'll never listen to you about anything ever again," wrote another. Health professionals, some of whom I used to respect, also challenged me – not with facts, but rather rants about how misguided and wrong I was about this issue.

If the goal was to stop my investigation, or to keep me from posting more videos, it failed miserably. Instead I became even more curious. Where were these people getting the idea that the flu was suddenly a threat to the survival of the human race? Why were people becoming unhinged about COVID-19?

As the entire world started shutting down, my COVID-19 investigation became practically a full-time job for an exceptionally long time. Instead of two videos per week, I started posting four. Instead of a few hundred responses per week, I started receiving over a thousand per day. People sent me documents and videos from all over the world that revealed that something awful was going on. It was not the flu, but it spread just as quickly as the flu, and was far more dangerous than any virus. It was a planned and carefully orchestrated event involving a campaign of misinformation enabled by a willing and obedient media that allowed some bad actors to gain control over billions of people, and to shut down the world's economies. Curiosity turned into shock and horror, and I decided that the entire story needed to be told.

The "COVID-19 Hoax," as we have always referred to it, is the most consequential thing that has happened in our world since World War II. But it is far worse than the war. WWII was caused by a lunatic in Germany with a relatively small following. Hitler and his followers inflicted what was unprecedented damage on the world, but they were identifiable enemies who could be overcome.

The COVID-19 Hoax has been perpetrated by a wide network of enemies of the people who have managed to disguise themselves as public servants, health professionals, and founders and heads of global non-profit organizations. They are everywhere, they are incredibly rich, they are powerful, and they are intent on getting their way. Overcoming these enemies will not be easy.

The human race is not in danger of extinction from COVID-19. Humans – all of us – are in danger of having our lives controlled by these enemies of the people who took our liberties and freedoms away in a matter of weeks. They have already shown that they plan to determine the circumstances under which we will be able to leave our homes, to interact with others, and to operate our businesses. In a very short time, they were able to destroy our communities, eliminate freedom of religion and the right to assemble, and turn us against each other. They are remarkably effective at their game. We cannot let them win.

This book tells the story of how this debacle began, who the major players are, and how the plan was executed. It's a disturbing account of how easy it is to take control of people who assume they will continue to enjoy freedom without holding government, the media, and many others accountable, and without any effort on their part.

With knowledge comes power, and our hope is that the information in this book will motivate masses of people to reject what is often referred to as "the new normal," to do whatever is required to take back their lives, and to ensure that future generations will live free.

> "As long as we have faith in our own cause and
> an unconquerable will, victory will not be denied us."
>
> —Winston Churchill

INTRODUCTION

by co-author Shane Prier

WITHIN A SHORT time after the COVID-19 hysteria began, I knew something was not right, and I started to investigate. I have a degree in biology and spent some time working in a lab researching viruses, bacteria, DNA, electrophoresis, and polymerase chain reaction technology. I know from experience that while viruses can be dangerous and lethal, they usually aren't. COVID-19 is a coronavirus, and most of us have been exposed to one or more coronaviruses during our lifetimes since this virus is the cause of what we refer to as "the common cold." Based on what I knew about coronaviruses, along with an analysis of the daily reports of cases, the actions taken by health authorities in response seemed exaggerated. And over time the disconnect between the data and the rules and regulations imposed on the entire world's population became more and more frightening. Something else had to be going on.

I also have a background in finance. For the last seventeen years, I've worked as a private wealth advisor and managed over

$100,000,000 in assets for my clients. I'm trained to look at almost everything that goes on in the world, and to evaluate it in terms of what it means to both the financial markets and politics because both have a significant impact on financial stability and wealth. It soon seemed apparent to me that there were powerful financial and political incentives for staging a pandemic. It also became clear over time who the potential winners of this debacle would be.

In this section, I draw on my biology/science/research background to explain some basics about viruses, and also about herd immunity, which makes viruses a lot less frightening. Later in the book, you'll learn more about the financial and political ramifications of COVID-19 and how both have motivated many of the bad decisions that have been made and continue to be made in response to it.

What is a Virus?

According to most scientists, viruses have been around since life itself. There is some debate about whether viruses should be considered life, given the fact that they cannot propagate without a host cell, but that is an argument that has gone on for decades. There is also some debate about the original natural origins of viruses. Are they the result of discarded genetic material from cells, or did they come first? Again, this is a debate that continues.

So, what is a virus? To better explain this, it is best to start with a basic description of cells, deoxyribonucleic acid (DNA), and ribonucleic acid (RNA). Most of us are familiar with the concept of DNA. It is what gives us our features and characteristics. Whether you have big ears, a small nose, brown hair versus blonde hair, etc., this is where it comes from.

DNA is housed in the nucleus of our cells. DNA is a double stranded molecule, which is bound together by what are called nucleotides. There are four of them in DNA. They are called Adenine (A), Guanine (G), Cytosine (C), and Thymine (T). The two strands bind together because they complement each other in terms of their molecular properties. A always binds with T, C always binds with G.

DNA by itself, is not able to produce the proteins necessary to give us our characteristics or perform the necessary cellular functions. First, a process called transcription must occur. Transcription is the process, via an enzyme called RNA Polymerase, which creates a single stranded copy from the DNA. It is an identical copy from one of the DNA strands, with one exception. Thymine is replaced by Uracil (U). This single stranded copy is the RNA, which is then released from the nucleus into the cytoplasm of the cell, which is the open area surrounding the nucleus. From here, with the use of cell organelles called ribosomes and endoplasmic reticula, the process of translation occurs. Here is where the "magic" happens. The ribosome basically moves down the strand of RNA, reads its blueprint, and then directs the production of whatever amino acid (the building blocks of proteins) it tells it to do. A commonly used analogy is that of a plastic zip tie. You insert one end into the slot on the other end and tighten it. The process of moving the slot down the actual strand itself, is a great way to think of how a ribosome reads the blueprint of the RNA.

Viruses are strands of either DNA or RNA, that are protected by a protein coat, and in some cases, an additional lipid envelope, which is a layer of fatty molecules that protect the virus. Viruses are so small, that they can only be seen under an electron microscope. It is important to note that viruses are in our bodies all the time. They are always around us, in our environment, in the air, in fluids, etc. Most viruses are not harmful. However, viruses do mutate over time. In other words, their genetic code can change slightly. In some cases, this can make them harmful to our cells. They may evolve a sequence of proteins on the outer layer, that bind perfectly, like a lock and key, to the receptors on the outer membranes of our cells. The receptors are protein compounds, that bind only with certain other proteins. This is how the cell determines what is "good" and what is "bad" …what to let in. If a virus mutates to the point where it has "cracked the code", it can bind to these receptors. Once it is able to do that, it injects its DNA or RNA (depending on the virus) into the cell itself.

Once inside the cytoplasm of the cell, in the case of an RNA virus such as COVID-19, the cell immediately goes to work on replicating the virus. The reason for this, is because the cell does not recognize it as an unwelcome visitor. One could say that the virus hijacks the machinery of the cell and uses it against it. Once inside, the RNA polymerase gets to work, replicating the invading virus strand. This continues, and eventually, the copied virus strains are released from the cell with the opportunity to invade other cells in the same fashion. In some cases, the replication within the cell is so significant that it causes the cell to burst. The result is the same, the viral copies are released and can invade other cells.

This all may sound very grim, but there is good news. The human immune system is amazing. Initially our cells do not recognize that the virus is an invader, and yes, it is able to do some damage during that time. People who have health issues are less able to withstand the negative effects of the virus, and can even die as a result. But in healthy people, the immune system quickly figures out what to do and springs into action.

Eventually the invaded cells notice the chemical imbalances within and send out a "distress" call. This distress call is in the form of proteins which are sent to their own outer membrane. The first to respond are T-cells, a type of white blood cell. They are called this because they mature in the thyroid. T-cells look for distressed cells and then destroy them by either engulfing them, or by binding to the distress proteins and injecting what are called granzymes into the cell, which initiates a process called programmed cell death, or apoptosis. So, the distressed cell performs a very altruistic act by basically asking to be killed, for the benefit of the whole body, or community of cells of the organism. The goal of the destruction of infected cells is to stop the virus from replicating and spreading to other cells.

While T-cells are doing their job, the immune system mobilizes B-lymphocytes, another type of white blood cell, to produce proteins called antibodies. It takes a little longer to produce antibodies because they are custom-made for each foreign invader, or antigen. Antibodies bind to the virus and fit like a key fits a lock. The virus is either neutralized or destroyed. Once antibodies are produced for a particular virus, a person has immunity from the virus for at least a few years should the virus be encountered again.

If the virus is attached to a cell at the time, the lymphocyte will often initiate, using a lysin, a process called lysis, which is another form of cell death that involves the penetration of the cell membrane. Once neutralized, the T-cells recognize the protein sequences, and destroy the attached virus.

Some viruses are harder to stop than others. An example is Human Immunodeficiency Virus (HIV), which first appeared in the 1970s and 1980s, and remains a major problem in third world countries today. HIV is a more difficult infection to address because the virus actually attacks T-cells, part of the army of immune cells that are supposed to neutralize viruses.

But the virus that is the subject of this book is a coronavirus, and this family of viruses has been with humans for a very long time. Anyone who has ever had a cold, has been infected with a coronavirus. The name "coronavirus" is due to the appearance of these viruses when viewed under an electron microscope. They appear to have a crown, or a "corona." This is due to the fact that the outer protein coat contains "spikes" called S-Proteins that are integral to binding to cell receptors.

Coronaviruses are RNA viruses, which means that once inside the cell, replication can occur almost immediately. They are respiratory viruses like influenza, which means they can be airborne, and survive on surfaces for extended periods of time.

They enter through the eyes, nose and mouth, and attack the lung cells. This is different than a gastric virus, for example, which enters typically through the mouth and goes to the gastro-intestinal system, often when contaminated food is eaten. Because of the airborne characteristic, respiratory viruses tend to be more contagious than these others.

Taming the Herd

"When enough of us represent such 'dead ends' for viral transmission, spread through the population is blunted, and eventually terminated. This is called herd immunity."[1]

– Dr. David L. Katz,
President of True Health Initiative and founding
director of The Yale-Griffin Prevention Research Center

Herd immunity develops when a significant percentage of the population becomes infected with a virus and develops antibodies. These people will not become sick again if exposed to the virus again. This provides protection to people who are not immune to the disease because those who are immune will no longer spread the disease. it is thought that if 50%-70% of a population has become immune, the population has developed herd immunity, and the virus can no longer be spread, including to those people who have compromised immune systems.

Some researchers, like Gabriela Gomes, professor of mathematics and statistics at Strathclyde University, report that the threshold for herd immunity to COVID-19 is as low as 20%, and that we may already have reached it. This is due to the fact that so many people have had prior exposure to coronaviruses, and this provides some immunity.[2] Researchers at Oxford agree, and report that in addition to many people already achieving immunity, some people are just naturally resistant to all types of infections.[3]

Prior to COVID-19, placing a healthy person in quarantine was unheard of, and viruses were permitted to "run the course" until herd immunity was developed. Had healthy people been allowed to go about their daily lives during the COVID-19 health crisis, the amazing human body would have shown its stuff to yield protection for the individual, and avoid later spread to contacts with compromised immune systems. But health officials took unprecedented action and mandated that people "shelter at home."

This served to prevent the development of herd immunity and as you will learn later, was a terrible decision.

ENDNOTES

1. David L. Katz. Is Our Fight Against Coronavirus Worse Than the Disease? *New York Times.* March 20 2020 https://www.nytimes.com/2020/03/20/opinion/coronavirus-pandemic-social-distancing.html?action=click&module=RelatedLinks&pgtype=Article accessed 9.2.2020

2. James Hambiln A New Understanding of Herd Immunity. *The Atlantic* July 13 2020 https://www.theatlantic.com/health/archive/2020/07/herd-immunity-coronavirus/614035/ accessed 9.2.2020

3. Lourenco J, Pinotti F, Thompson C, Gupta S. "The impact of hot resistance on cumulative mortality and the threshold of herd immunity for SARS-CoV-2." *medRxiv* **doi:** https://doi.org/10.1101/2020.07.15.20154294

THE PERFECT STORM

LOOKING BACK OVER the course of many events during the last few decades, sooner or later something bad was bound to happen. Most likely no one (except the people who planned this) could have predicted that a made-up pandemic would be used to shut down the entire world. But many people have been saying that the state of education, medicine, government, and the media have been deteriorating for a very long time. As it turns out, they were right, and these factors created a perfect environment for the events that began in early 2020.

Let's start with education. Well-educated people know a lot about history, and also about the world they live in today. They are critical thinkers, and ask questions about what they read, see, and hear. Educated people can contribute to society in numerous ways, which include working in their communities, building strong families, and engaging in productive work. At one time, graduating from high school was an achievement worth working toward. A college education even more so because it led to higher lifetime earnings.

Things have changed, and not for the better. More than 30 million adults in the U.S. cannot read or write above the third grade level,[1] and 50% of adults can't read a book at the 8th grade level.[2] As of December 14, 2017, the U.S. Census Bureau reported that 90% of Americans age 25 and older had completed high school.[3] This means that most of the adults who cannot read at the eighth grade level are high school graduates. This is not good.

The situation is not much better for college graduates. In 2019, A Gallup-Lumina Foundation poll determined that only 13% of Americans thought that college graduates were well-prepared for successful employment. The sentiment was even stronger for Americans who had earned college degrees – only 6% of them thought that college grads were ready to start work successfully.[4] Colleges and universities are clearly not preparing students to lead productive and fulfilling lives.

The recent trend toward establishing "safe spaces" on college campuses provides an example of how ill-prepared many college students are for success in life. Safe spaces are places where students are protected from ideas and speech that has the potential to make them feel uncomfortable, or with which they disagree. The *New York Times* featured an article in 2015 on a safe space established at prestigious Brown University in response to on-campus talks about "the role of culture in sexual assault." Student volunteers advertised the room as a retreat for students to recuperate who found the debate "upsetting."

Emma Hall, a rape survivor and sexual assault peer educator, helped to set up the room, which was equipped with Play-Doh, calming music, pillows, blankets, cookies, coloring books, and a video featuring puppies playing. Emma reported attending part of the lecture, but had to return to the safe space because "I was feeling bombarded by a lot of viewpoints that really go against my dearly and closely held beliefs."[5] An inability to listen to anything other than speech that reinforces one's beliefs should be concerning to all of us.

But Brown's safe space was not an isolated case. By 2017, not only were safe spaces becoming more common on college campuses, but "controversial speakers," defined as individuals discussing a point of view that some students did not agree with, had their lectures cancelled. Some who showed up to deliver their talks were met with threats and violence and were forced to leave.

Colleges and universities were at one time places in which students learned how to listen to different points of view and if they disagreed, explain the basis for their disagreement. Open debate was part of college life. Protecting people from hearing opposing points of view does not lead to the ability to engage in critical or independent thinking.

Why is this important? It is easier to "sell" a false story to a population of people if a significant percentage of them cannot read or comprehend or think critically. As you will learn in this book, there were no data to support the declaration of a pandemic at the time it was announced, and as time went on, data clearly showed that the declaration was unwarranted. The stories about the "pandemic" became more and more implausible even as restrictions on the daily life of citizens became more onerous.

For example, small businesses were closed because being around other people was "dangerous" and to be avoided. Buying food at crowded grocery stores and purchasing goods at big box stores, however, was not dangerous. In other words, the virus was somehow neutralized while a person shopped at Wal Mart but became quite virulent in a small shoe store. Who would believe such nonsense? People who are poorly educated, and who have not been taught to engage in independent and critical thinking.

Medicine has a long history of recklessness and arrogance and has been careening toward disaster for a long time. Many books have been written about this topic, but perhaps the most important issue to discuss here is how easily doctors and the medical profession can be persuaded to do just about anything. The best example – the recent opioid epidemic which should have led to major reforms in medical training and practice but did not. Many people still think that the illegal "pill mills" were the main cause of the opioid crisis, but they played a rather small role. It was organized medicine, multi-billion dollar drug companies, and government partners that caused this disaster. The story of what really happened provides some insight as to how the COVID-19 debacle could be orchestrated.

While there were many manufacturers of opioid drugs, one of the biggest and most influential was Purdue Pharma. The President's Commission on Combating Drug Addiction and the Opioid Crisis concluded that Purdue Pharma's marketing program and the company's investment of billions of dollars in influencing government regulation and medical policy resulted in a 10-fold increase in opioid prescribing in the U.S. Purdue benefitted greatly. Several members of the Sackler family, principal shareholders in Purdue and a few other opioid manufacturing companies, became billionaires as a result. Purdue was not the only drug company that profited from the opioid epidemic. How did this happen?

Bertha Madras of Harvard Medical School was principal author of the Commission's report.[6] She says the drug companies invested enormous amounts of money that "literally bought off" the Joint Commission (which accredits hospitals and sets medical policies), the Federation of State Medical Boards, several American pain associations, and the legislature, by investing almost $2.5 billion in both lobbying and funding members of Congress. In fact, there are more drug industry lobbyists than elected members of Congress, which allowed the companies to successfully block the efforts of a few legislators that knew there was a problem and tried to do something about it. Drug companies even controlled physician training and the development of guidelines for treating pain. In effect, the drug companies corrupted the very institutions that should have been protecting Americans from them, converting them from regulators to business partners.

The investment paid off. By 2012, American doctors were writing 255 million prescriptions per year for opioid drugs.[7]

The drug companies were well aware of the negative impact of their products and actively pursued strategies to expand the market for them anyway. For example, Johnson and Johnson hired the prestigious consulting firm McKinsey and Company to help increase sales. McKinsey recommended targeting doctors who were prescribing large amounts of OxyContin and advised the drug maker to "target high abuse-risk patients." This sounds more like conversations that might take place between members of a drug cartel than what one would expect from strategy sessions involving American business executives.

McKinsey also served as a consultant to Purdue, advising both Purdue and Johnson and Johnson to "invent" an epidemic of untreated pain to increase demand for their products. This turned out to be a successful strategy. Makers of opioids funded what appeared to be an independent organization called the American Pain Society (APS), which promoted the idea that pain relief with opioids was a human right. The APS was responsible for convincing the Veterans Administration and the Joint Commission (TJC, formerly The Joint Commission on the Accreditation of Healthcare Organizations or JCAHO) to recognize pain as the "fifth vital sign," along with markers like blood pressure and temperature, and to prioritize treatment of pain with opioids.

A lawsuit filed by several municipalities in West Virginia against TJC reveals that TJC partnered with Purdue Pharma and other opioid makers to issue Pain Management Standards that misrepresented the risk of opioid addiction and resulted in inappropriate prescribing of the drugs. The lawsuit alleges that TJC continued this partnership even after Purdue had pleaded guilty to felony criminal charges for misrepresentation concerning OxyContin in 2007.[8]

According to Chris McGreal, author of *American Overdose,* Purdue wrote and distributed "educational materials" for free for TJC in return for opportunities to interact with and train medical staff. Videos and manuals stated that concerns about addiction and overdose were "inaccurate and exaggerated." Additionally, Purdue funded over 20,000 educational programs on pain which were thinly disguised sales seminars for the company's products.

On TJC's website, the organization disingenuously presents itself as an independent organization with a mission "To continuously improve health care for the public."[9] TJC certifies over 22,000 healthcare organizations and programs in the U.S., issues a Gold Seal® of Approval to qualifying institutions, and states that its vision is for all people to experience the safest and best quality healthcare.[10] This seems inconsistent with the organization's actions, which included partnership with Purdue to expand the prescribing of opioid drugs.

TJC is a powerful organization, and its standards dictate the way American hospitals and medical facilities are operated. It forced medical institutions and healthcare professionals to actively look for pain in patients and treat it with opioids. According to McGreal, the Joint Commission has recently changed its guidelines, but denies any wrongdoing. The Joint Commission states that doctors and the APS were to blame for the opioid crisis because they presented false evidence. This is interesting – how can the TJC claim to offer a certification that has any meaning at all if it essentially believes any information presented to it without further investigation? The APS disbanded in 2019, claiming it was the victim of a witch hunt. None of the groups or individuals involved in this debacle seem to have any intention of taking responsibility and instead present themselves as victims.

The Federation of State Medical Boards is a nonprofit organization that develops guidelines for 70 state medical boards in the U.S. and its territories, and co-sponsors medical licensing examinations. The Federation took $100,000 from Purdue Pharma to help pay for the printing and distribution of "Responsible Opioid Prescribing: A Physician's Guide." The Federation estimated that it would need $3 million dollars to complete its marketing program to promote the safe use of opioid drugs for chronic pain. Six other makers of opioids were asked to contribute to this campaign."[11]

The FDA is equally culpable, approving new opioid drugs while more and more Americans were dying because of taking them. The agency is still doing this. It approved Dsuvia® in 2018,[12] a more potent version of fentanyl. This decision, which is difficult to reconcile in view of the ongoing opioid crisis, results from the fact that the FDA is funded primarily by the drug companies, and now collects about $2.6 billion dollars annually from industry.[13] Not surprisingly, the approval rate for drugs is 96%.[14] Even though a significant percentage of its operating budget comes from drug companies in the form of these fees, the FDA denies that this financial support influences its decisions. Right.

The American Medical Association (AMA) played a role too, opposing a law introduced in Congress that would have required doctors to receive training to prescribe opioids. Members of Congress funded by Big Pharma helped the AMA to kill the bill, with public attacks on sponsors and advocates. Congressman Butterfield, a Democrat from North Carolina, praised the drug distributors for their "very impressive" efforts to stop opioids from ending up in the hands of people who should not take them. The irony was that at the same time that this statement was made, the companies were paying fines to the Justice Department for failing to report suspicious orders from small rural pharmacies for millions of pills.[15]

Why is this important?

Orchestrating a fake pandemic required that medical professionals "buy in" to the idea that there actually was a pandemic and that they remain on board even as the data started to clearly show that something was amiss. The World Health Organization (WHO) characterized the COVID-19 outbreak as a pandemic on March 11, 2020, when there were 118,000 cases and 4291 people had died worldwide.[16]

By March 26, the World Health Organization reported that there were 575,444 cases and 26,654 deaths from coronavirus since this debacle began.[17] According to the Centers for Disease Control (CDC), as of March 28 there were 103,321 cases and 1668 deaths from coronavirus in the U.S.[18]

Another CDC website reported that during the 2018-2019 flu season, there were an estimated 35.5 million cases, 406,600 hospitalizations, and 34,200 deaths from seasonal flu.[19] The CDC estimated that for the 2019-2020 season, there would be between 17 million and 24 million medical visits for flu in the U.S., 370,000-670,000 hospitalizations for flu, and between 22,000 and 55,000 deaths.[20]

These data, which were taken directly from the WHO and CDC websites, simply do not make sense. How can 575,444 cases of coronavirus worldwide result in the world as we know it slowly being shut down, while tens of millions of cases and as many as 13-33 times the number of deaths from seasonal flu in the U.S. alone require no response at all (with the exception of constant nagging to get a flu shot, which the CDC acknowledges may or may not work).[21]

Most directors of state health departments quickly bought the story. It is apparent that none of them actually checked to confirm that the declaration of a pandemic was substantiated. It did not take much to convince these doctors that the hospitals would soon be overwhelmed, and that hundreds of thousands, if not millions of people were going to die. The lockdown orders started.

But what was truly amazing is that weeks and months after the pandemic failed to materialize – the hospitals were empty and the death rate was lower than seasonal flu - these doctors continued to behave as if COVID-19 was still an imminent threat.

The people who orchestrated this debacle knew that doctors are easy to convince of almost – well, apparently, anything. They could count on the medical profession to implement the plan regardless of what the data showed.

The media, at one time, was comprised of principled reporters and investigators who were determined to find and report the truth about government, politicians, businesses, and important issues of the day. In recent years, however, the mainstream media has degenerated, and ethical decision-making and journalistic integrity seem to be things of the past. This view is widespread, and polls have consistently shown that most people no longer trust the media. One survey showed that inaccuracy, bias, alternative facts, and too much reporting based on opinions and emotions were reasons for distrusting the media.[22]

Why is this important?

The plan required that reporters and media outlets dutifully report any information given to them. Fact-checking and asking challenging questions would turn up inconvenient information that might cause people to question whether there was a pandemic, and if the responses to it were reasonable. Powerful people and institutions knew that the media could be counted on to deliver a carefully crafted message and to induce and maintain a panicked population by repeating the same inaccurate messages again and again.

Governments in westernized countries have expanded exponentially and exerted increasing levels of control over citizens. The conversion of elected officials to rulers of people began a long time ago and was rather gradual, so as not to alarm most people.

The idea of increased government control had become so mainstream that by early 2020, Bernie Sanders, who openly praised Fidel Castro[23] and Communist China,[24] was on his way to earning the Democratic nomination for President of the United States. The party did not let this happen. But this showed that many U.S. citizens thought at the time that socialism and communism, both of which involve incredible levels of government control, were acceptable.

Why is this important?

It is easier to orchestrate a fake pandemic if significant numbers of citizens can be counted on to do as they are told by government officials. Instructions included sheltering in place, closing businesses, wearing masks (sometimes even when inside one's own home), consenting to temperature checks and other medical evaluation at hair salons and airports, and snitching on people who violated the orders, sometimes even family members and friends.

The bottom line: The timing for COVID Operation was perfect. An uneducated populace, a significant percentage who favored more government control; government officials, many elected because they promised to institute more big government plans; medical doctors who could be convinced to do almost anything; and a media that would promote and reinforce the narrative were the necessary ingredients to pull off the biggest hoax in the history of the world.

ENDNOTES

1. Editorial Team. Crisis Point: The State of Literacy in America. *Resilient Educator.* https://resilienteducator.com/news/illiteracy-in-america/. Accessed June 6, 2020.

2. Valerie Strauss. Hiding in plain sight. The adult literacy crisis. *Washington Post.* November 1, 2016.

3. United States Census. High School Completion Rate Is Highest in U.S. History. December 14, 2017. https://www.census.gov/newsroom/press-releases/2017/educational-attainment-2017.html#:~:text=Dec.,from%20the%20U.S.%20Census%20Bureau. Accessed June 6, 2020.

4. Brandon Busteed. America's "No Confidence" Vote on College Grads' Work on Readiness. Gallup. April 24, 2015, https://news.gallup.com/opinion/gallup/182867/america-no-confidence-vote-college-grads-work-readiness.aspx, Accessed June 8, 2020.

5. Judith Shulevitz. In College and Hiding From Scary Ideas. *New York Times.* March 21, 2015.

6. President's Commission on Opioids. The White House. https://www.whitehouse.gov/ondcp/the-administrations-approach/presidents-commission-opioids/. Accessed August 15, 2020.

7. Centers for Disease Control and Prevention. U.S. Opioid Prescribing Rate Maps. Centers for Disease Control and Prevention. https://www.cdc.gov/drugoverdose/maps/rxrate-maps.html. Accessed August 15, 2020.

8. City of Charleston West Virginia v. The Joint Commission. Unpublished. (S.D. WV 2020) https://www.statnews.com/wp-content/uploads/2017/11/2017-11-02-FINAL-Kenova-v-JCAHO-class-action-complaint.pdf

9. The Joint Commission. https://www.jointcommission.org/. Accessed August 15, 2020.

10. Facts About The Joint Commission. The Joint Commission. https://www.jointcommission.org/facts_about_the_joint_commission/.

11. John Fauber. Follow the Money: Pain, Policy and Profit. *MedPage Today.* https://www.medpagetoday.com/neurology/painmanagement/31256. February 19, 2012.

12. Drug Approval Package: DSUVIA. U.S. Food & Drug Administration. https://www.accessdata.fda.gov/drugsatfda_docs/nda/2018/209128Orig1s000TOC.cfm.

13. Fact Sheet: FDA at a Glance. U.S. Food & Drug Administration. https://www.fda.gov/about-fda/fda-basics/fact-sheet-fda-glance. Accessed August 15, 2020.

14. Mathew Herper. The FDA Is Basically Approving Everything. Here's The Data To Prove It. *Forbes.*

15. http://www.forbes.com/sites/matthewherper/2015/08/20/the-fda-is-basically-approving-everything-heres-the-data-to-prove-it/.

16. Chris McGreal. Big Pharma's Response to the Opioid Epidemic: Pay But Deny. *NYR Daily.* https://www.nybooks.com/daily/2019/11/11/big-pharmas-response-to-the-opioid-epidemic-pay-but-deny/.

17. WHO Director-General's opening remarks at the media briefing on COVID-19 - 11 March 2020. World Health Organization. https://www.who.int/dg/speeches/detail/who-director-general-s-opening-remarks-at-the-media-briefing-on-covid-19---11-march-2020. Accessed June 6, 2020.

18. Coronavirus disease (COVID-19) pandemic. World Health Organization. https://www.who.int/emergencies/diseases/novel-coronavirus-2019.

19. Coronavirus Disease 2019 (COVID-19): Cases in the US. Centers for Disease Control and Prevention. https://www.cdc.gov/coronavirus/2019-ncov/cases-updates/cases-in-us.html.

20. Estimated Influenza Illness, Medical visits, Hospitalizations, and Deaths in the United States – 2018-2019 influenza season. Centers for Disease Control and Prevention. https://www.cdc.gov/flu/about/burden/2018-2019.html.

21. Preliminary In-Season 2019-2020 Burden Estimates. Centers for Disease Control and Prevention. https://www.cdc.gov/flu/about/burden/preliminary-in-season-estimates.htm.

22. Ibid.

23. Indicators of Media Trust. Knight Foundation. https://knightfoundation.org/reports/indicators-of-news-media-trust/. Accessed June 6, 2020.

24. Kenneth Garger. Bernie Sanders defends Fidel Castro's socialist Cuba: 'Unfair to simply say everything is bad'. *New York Post.* February 23, 2020.

25. Joseph A. Wulfsohn. Bernie Sanders doubles down on Fidel Castro praise: 'Truth is truth'. *Fox News.* February 25, 2020.

A PREVIEW OF THINGS TO COME

ACCORDING TO the Centers for Disease Control's (CDC) website, about 9% of the world's population is affected by the seasonal flu annually with up to one billion infections, 3-5 million severe cases, and 300,000-500,000 deaths per year.[1] [2] It is estimated that 20% of Americans are affected, with 25-50 million documented cases, 225,000 hospitalizations and tens of thousands of deaths annually.[3] [4] [5] [6] [7] Historically, the elderly account for 90% of influenza deaths.[8] These data are for "normal" years.

One season that was considered abnormal was 2009-2010, during which the swine flu (H1N1) was circulating. CDC data shows that there were 60.8 million cases, 274,304 hospitalizations, and 12,469 deaths from swine flu in the U.S.[9] It is estimated that as many as 575,400 people died of H1N1 worldwide during a one-year period.[10]

On June 11, 2009 the World Health Organization declared that H1N1 was a pandemic after 70 countries reported cases of novel influenza A (H1N1).[11] Records show that even with the higher number of cases and deaths with H1N1, there was no reason to declare a pandemic.

A sign that something was not right was when the CDC instructed health care professionals to stop testing for H1N1 and to assume that everyone who presented with flu-like symptoms had H1N1 flu and to report it as such. CDC's statement to the public was that resources should not be wasted on testing when the government had already determined that there was a pandemic. A possible reason for this policy change was that the U.S. government had ordered 193 million doses of flu vaccine and needed to "sell" these doses to the public; and the CDC's directive to stop testing was an attempt to prevent the public from learning the truth about the number of cases which most certainly would have reduced interest in the vaccine.

A CBS news investigation confirmed this was the case. CBS requested state-by-state testing results for the period prior to the halting of lab testing by the CDC. The CDC refused to provide the data, so CBS filed a request under the Freedom of Information Act with the Department of Health and Human Services, and also asked all 50 states to provide their data on lab-confirmed H1N1 prior to the order to discontinue testing.

The results when the data were finally turned over? The majority of patients were negative for both H1N1 and seasonal flu even though the states were testing patients deemed to be at the highest risk of having H1N1, such as people who had visited Mexico. Health authorities reported that these people had colds or upper respiratory infections, but not flu.[12]

The bottom line was that the predicted epidemic of flu, along with deaths and co-morbidity did not take place. Instead of telling the public the truth, the CDC lied to make its prediction appear to be true, and to promote flu vaccines.[13]

The World Health Organization was an active participant in the deception. According to the Committee on Social, Health and Family Affairs of the Parliamentary Assembly of the Council of Europe (PACE), the WHO engaged in fear-mongering with H1N1 flu, without any evidence to back its actions. As a result, approximately $18 billion dollars was squandered worldwide.

PACE determined WHO colluded with the drug companies, turning "run-of-the-mill" flu into a "pandemic." Drug companies benefitted financially, millions of people were vaccinated without cause, and without evidence that the vaccine was effective since it was not clinically tested. Testimony at a public hearing included this statement, "We are witnessing a gigantic misallocation of resources in terms of public health. Governments and public health services are wasting huge amounts of money in investing in pandemic diseases whose evidence base is weak."[14]

Epidemiologist Wolfgang Wodarg said the "false pandemic" is one of the greatest medical scandals of the century.[15] Dr. Ulrich Kiel, director of the WHO Collaborating Center for Epidemiology in Munster Germany agreed.[16] The swine flu was actually much milder than seasonal flu. The CDC later reported that the death rate was one tenth to one third of the death rate from normal seasonal flu.[17]

Why would the WHO have declared a pandemic when there was none? Not surprisingly, the drug companies influenced the response with so much to gain. Three scientists who helped to develop WHO guidance on pandemic influenza preparedness had consulted for drug companies that would profit from WHO policies concerning pandemics. These conflicts were not disclosed in a document titled *WHO Guidelines on the Use of Antivirals and Vaccines During an Influenza Pandemic* issued in 2004. The WHO refused to reveal the names of the members of the WHO Emergency Committee set up to guide response to the H1N1 "pandemic." This is concerning since the responsibility of this committee included when to change/increase the response rate.[18]

An additional motivation was revealed by Margaret Chan, at the time WHO Director-General. She said in a speech that "ministers of health" should take advantage of the "devastating impact" swine flu will have on poorer nations to get out the message that "changes in the functioning of the global economy" are needed to "distribute wealth on the basis of" values "like community, solidarity, equity and social justice." She further declared that the pandemic should be used as a weapon against "international policies and systems that govern financial markets, economies, commerce, trade, and foreign affairs."[19] In other words, Chan looked at fake pandemics as a form of social engineering, to be executed according to her beliefs, of course.

Other than some temporary embarrassment, there were no real consequences for any of the people who were active participants in the swine flu deception. So, there would be no reason not to do the same thing again in the future.

20

ENDNOTES

1. Lambert LC, Fauci AS. Influenza Vaccines for the future. *NEJM* 2010;363(21):2036-2044.

2. Centers for Disease Control and Prevention. Estimates of deaths associated with seasonal influenza – United States, 1976-2007. *MMWR Morb Mortal Wkly Rep.* 2010;59(33):1057-1062. https://www.cdc.gov/mmwr/preview/mmwrhtml/mm5933a1.htm. Accessed August 27, 2020.

3. Lambert LC, Fauci AS. Influenza Vaccines for the future. *NEMJ* 2010;363(21):2036-2044.

4. Centers for Disease Control and Prevention. Estimates of deaths associated with seasonal influenza – United States, 1976-2007. *MMWR Morb Mortal Wkly Rep.* 2010;59(33):1057-1062. https://www.cdc.gov/mmwr/preview/mmwrhtml/mm5933a1.htm. Accessed August 27, 2020.

5. Simonsen L, Clarke MJ, Williamson GD, Stroup DF, Arden NH, Schonberger LB. The impact of influenza epidemics on mortality: introducing a severity index. *Am J Public Health* 1997;87(12):1944-1950.

6. Simonsen L, Fukuda K, Schonberger LB, Cox NJ. The impact of influenza epidemics on hospitalizations. *J Infect Dis* 2000;181(3):831-837.

7. Thompson WW, Shay DK, Weintraub W, et al. Influenza-associated hospitalizations in the United States. *JAMA* 2004;292(11):1333-1340..

8. Molinari NA, Ortega-Sanchez IR, Messonnier ML, et al. The annual impact of seasonal influenza in the US: measuring disease burden and costs. *Vaccine.* 2007;25(27):5086-5096.

9. 2009 H1N1 Pandemic. Centers for Disease Control and Prevention. https://www.cdc.gov/flu/pandemic-resources/ 2009-h1n1-pandemic.html. Accessed August 27, 2020.

10. Ibid.

11. Centers for Disease Control and Prevention. WHO Pandemic Declaration. https://www.cdc.gov/h1n1flu/who/. Accessed August 27, 2020.

12. Sharyl Attkisson. Swine Flu Cases Overestimated? October 21, 2009. CBS.

13. http://www.cbsnews.com/news/swine-flu-cases-overestimated/. Accessed August 27, 2020.

14. World Health Organization. Changes in reporting requirements for pandemic (H1N1) 2009 virus infection. July 16, 2009. http://www.who.int/csr/disease/swineflu/notes/h1n1_surveillance_20090710/en/. Accessed August 27, 2020.

15. Parliamentary Assembly. Extracts of statements made by the leading participants at the public hearing on "The handling of the H1N1 pandemic: more transparency needed?", organized by the Committee on Social, Health and Family Affairs of the Parliamentary Assembly of the Council of Europe (PACE) in Strasbourg on Tuesday 26 January 2010. http://assembly.coe.int/nw/xml/News/FeaturesManager-View-EN.asp?ID=900. Accessed August 27, 2020.

16. Michael Ash. H1N1 – Questions About Profit Incentive. Clinical Education. https://www.clinicaleducation.org/news/h1n1-questions-about-profit-incentive/. Accessed August 27, 2020.

17. Parliamentary Assembly. PACE to prepare report on the handling of the Swine Flu pandemic. http://assembly.coe.int/nw/xml/News/News-View-EN.asp?newsid=2724&lang=2. Accessed August 27, 2020.

18. Centers for Disease Control and Prevention. Weekly 2009 H1N1 Flu Media Briefing: Thursday, December 10, 2009 – 1:00PM. https://www.cdc.gov/media/transcripts/2009/t091210.htm. Accessed August 27, 2020.

19. Cohen D, Carter P. WHO and the pandemic flu "conspiracies." *BMJ* 2010;340:c2912.

20. Margaret Chan. Address to the Regional Committee for Europe (59th Session). World Health Organization. September 15, 2009. https://www.who.int/dg/speeches/2009/euro_regional_committee_20090815/en/. Accessed August 27, 2020.

MEET SOME OF THE KEY PLAYERS

THE COVID-19 debacle was, in part, directed by a group of people who had considerable prior experience working with one another. During their careers, and often with one another, they had made decisions and engaged in activities that were concerning to the media, government officials, and the public. Yet as you will see, there were no real consequences for their actions.

Get to Know Tedros Adhanom Ghebreyesus

Tedros, as he is called, hails from Ethiopia, where he held various government positions before being named Director-General of the World Health Organization.

Ethiopia is an ethnically diverse country, with over 100 languages spoken. The key ethnicities are Oromo, Amhara, Somali, and Tigray; these constitute over 75% of the population.

An ethnic Tigray, Tedros became a member of the Tigray People's Liberation Front (TPLF), which gained power after overthrowing a regime that was associated with the Amhara in 1991. The Tigrays constitute only 6% of the population but hold most of the political power in the country. TPLF was classified as a terrorist organization by the Terrorism Research and Analysis Consortium.

Tedros was hired by the Ethiopian Health Ministry and eventually became Health Minister in 2005. His tenure was marred by scandal. According to Human Rights Watch, TPLF engaged in "systematic discrimination and human rights abuses" by refusing to deliver healthcare to the Amhara because they were affiliated with the opposition party.[1] "The Ethiopian government is routinely using access to aid as a weapon to control people and crush dissent," said Rona Peligal, Africa director at Human Rights Watch. "If you don't play the ruling party's game, you get shut out. Yet foreign donors are rewarding this behavior with ever-larger sums of development aid."

Even more concerning was a report by census committee chair Saima Zekaria showing that at least 2.5 million and possibly 6 million Amharas had disappeared from the census. According to Samira, the committee analyzed the data again and again and even brought in outside experts and could not find the reason for the discrepancy.[2] She stated publicly that there was an "obvious systematic reduction in the number of Amharas."[3]

During the time Tedros was health minister, there were three major cholera outbreaks in Ethiopia – in 2006, 2009, and 2011. He responded by reclassifying the outbreaks as acute watery diarrhea, which he did to avoid international embarrassment.

Outside experts who tested stool samples found cholera bacteria.[4] According to United Nations officials, more aid would have been delivered had Tedros told the truth about the outbreaks.[5]

An Audit Report from The Global Fund to Fight AIDS, Tuberculosis, and Malaria showed significant deficiencies in the way funds were managed by Tedros. The Office of the Inspector General visited 77 newly constructed health centers and found that 71% did not have access to water, 32% did not have functioning toilet facilities, 53% had major cracks in the floors and 19% had leaking roofs. Only 14% of the centers had equipment like microscopes and delivery beds, 12% had functional pharmacies, and none of the laboratories had hard work surfaces. Furthermore, millions of dollars of ineligible expenses were charged to grant programs, financial records were kept on Excel spread sheets, bank statements were not reconciled, and disbursements were not regularly reviewed.[6]

In 2012, Tedros was appointed Foreign Minister of Ethiopia. During his tenure in this position, he failed to act when several foreign countries, including Saudi Arabia, South Africa, and Yemen, repatriated thousands of Ethiopians who had moved to these countries due to the high unemployment rate in Ethiopia. While other countries like Kenya and Nigeria played an active role in helping their citizens return home, Tedros watched and did nothing while thousands of his citizens were imprisoned, beaten, and killed.[7]

As soon as Tedros was nominated for the position of Director-General of the World Health Organization, almost all articles and reports of his failings as Health Minister and Foreign Minister were purged. His healthcare credentials were amplified to focus on his successes and conveniently omitted his failures. Surprisingly, as of the time this book was going to press, a full-length documentary remained online which featured screen shots of articles and still shots from videos describing the carnage that took place during Tedros' time in office.[8]

Based on this information, Tedros would seem like a strange choice for Director-General. As you will see later, he has friends in high places.

It is said that leopards do not change their spots, and so it is with Tedros. Shortly after becoming head of the WHO, Tedros appointed Robert Mugabe, former president of Zimbabwe as goodwill ambassador to the WHO. Mugabe's regime was known for violence, and he was known as a ruthless ruler who once declared that only God could remove him from office.[9] Due to public pressure, the offer was withdrawn.

Bill Gates: The Power and Danger of Having Too Much Money

From an early age, Bill Gates was known to be highly intelligent, at least from a mathematical standpoint. He told a teacher that he would be a millionaire by age 30. Interestingly, how he acquired money and power is not well-known. What many believe, and what Bill Gates tells the world, is that he is "self-made." The truth of the matter is that he had help – a lot of help. Gates was born into a wealthy family.[10] His father was a successful attorney, and his mother, Mary, was a prominent businesswoman who used her influence and made important introductions that paved the way for young Bill to succeed. She was the first woman to lead United Way of America, and John Opel, chairman of IBM, served with her on the executive committee. Following a conversation between Mary and Opel, IBM hired Microsoft to develop an operating system for its soon-to-be-introduced first personal computer. This was how Microsoft became the world's leading software maker.[11]

Knowing Bill Gates' personal history, and his family's influence is necessary in understanding his role in the coronavirus debacle. Gates says that his father (Bill Gates Sr.) was the most influential person in his life. His father's history provides some insight into how Bill Jr. became a power-hungry globalist who states that population control is one of his primary goals.

Bill Gates Sr. sat on the board of Planned Parenthood,[12] an organization that takes in most of its revenue from abortions and taxpayer-funded federal money. The founder of planned parenthood, Margaret Sanger, was a self-proclaimed advocate of eugenics and population control and was openly racist, as evidenced by her speech at a Ku Klux Klan rally for women.[13] Planned Parenthood clinics are primarily located in African American neighborhoods, and the organization provides a significant percentage of abortions performed in the U.S.[14] Gates Sr. was also close with the Rockefeller and Soros families. The Rockefeller Foundation has received millions in donations from The Bill and Melinda Gates Foundation (BMGF),[15] strongly advocates for population control,[16] and has openly stated for years that the foundation is interested in promoting worldwide vaccination[17] and population control.[18]

Microsoft had significant legal problems under young Bill's management. The company was sued multiple times for anti-trust violations. The U.S. government ultimately decided that Microsoft had created an illegal monopoly on operating systems and their components in order to limit competition from other companies, and to inflate prices for these products.[19] Microsoft products were the subject of considerable criticism over the years. According to Dan Kusnetzky, president of the Kusnetzky Group, a prestigious information technology (IT) industry research firm, many companies developed products that were better than Windows but could not sell their products because of the stranglehold Microsoft maintained on the industry."[20] During antitrust depositions, Bill Gates came across as arrogant and stubborn, and unwilling to admit that there was anything wrong with his products or his business practices.

Gates behaved in much the same way as he built his foundation and changed his focus to healthcare. Believing he is smarter than everyone else, he has used his stubborn, arrogant disposition and monopolistic behavior to take over the world's healthcare systems. With unlimited funds to invest in getting the right people to see things his way, it is worth looking at his motivation, agenda, and the products and services he and the foundation promote – primarily population control and vaccinations.

At a TED conference in 2010, Bill Gates said "Now if we do a really great job on new vaccines, health care, reproductive health services, we lower that [the population] by perhaps 10 or 15 percent."[21] According to the BMGF website, the foundation has given over $11 million to the Population Council between 2018 and 2020, a group dedicated to population control.[22] The Council was founded by John D. Rockefeller in 1952 and the organization's website states that population growth is a barrier to social and economic development in developing countries; and that increased longevity in developed countries increase the cost of social welfare programs. The council conducts research on the effect of population growth on societies, families, and individuals, which seems to go beyond just "family planning," which it promotes as its primary activity.[23]

Like his parents, Gates seems to have interesting friends, including the late Jeffrey Epstein. Believing that his DNA was superior to that of others, Epstein was a eugenicist who planned to impregnate young girls at a ranch he owned in New Mexico to create a superior race. Bill Gates met with Epstein several times over the years, even after he was convicted as a sex offender.[24] Gates flew on Epstein's private jet, although later he claimed he did not know it was his. When Epstein was found dead from apparent suicide on August 10, 2019, Gates may have been one of many prominent wealthy people who were relieved that the sex offender and eugenics promoter would not be able to speak publicly about his relationships with them.

Both Bill and Melinda Gates are enthusiastic advocates for vaccinations and have publicly put forth an interesting theory about how vaccines can help to control population growth. They say that the reason women have so many children in third world and developing countries is that the infant and childhood death rate is so high, which leads to having more and more children in order to compensate. With vaccines saving the lives of children, they say, there will not be the need to have so many, and the population will ultimately be reduced. It is an unproven theory, perhaps because the math does not work out so well. If births of more children are to replace dead children, it would not appear that the population is growing due to too many children born in each family. The inability to explain this to anyone has apparently not stopped the Gates Foundation from using vaccines for the purpose of population control, but in a much different way.

Major pharmaceutical companies funded by the BMGF have vaccinated young women in Africa with compounds tainted with Human Chorionic Gonadotropin (HCG), which causes miscarriages for pregnant women and sterilization for women who are not pregnant.[25] This vaccination program was promoted by the World Health Organization, and pregnant women were told that the vaccine would prevent tetanus in their babies when in fact, the babies would never be born.[26]

Based on suspicions of Catholic doctors working in Africa, three independent Nairobi biochemistry labs tested samples from a tetanus vaccine being used and HCG was found in half of them.[27]

In January of 2019, Obianuju Ekeocha (Uju), founder of "Culture of Life Africa" stated "I founded Culture of Life Africa in 2013, only a few months after the Gates Foundation stepped up their population control efforts. They call it 'family planning' in the developing world, but from my view it was a bold, audacious step towards population control."[28] As one who grew up and lived in Africa most of her life, she is appalled about how BMGF and other wealthy elites from the west are trying to control and depopulate the African continent. The thought of contraception and especially abortion is not even in the vocabulary of the African people, as she mentioned in an address to the United Nations.

In multiple interviews Uju has said "Africans don't want this!" In August of 2012, Uju wrote what she called an "Open Letter to Melinda Gates" stating the concerns of the African people regarding forced contraception, abortions, sterilizations, and other depopulation methods.[29] In her letter, she asked that BMGF instead provide goods and services that are much needed in Africa such as nutrition programs and food for children, and better neo-natal care. Not surprising that Melinda Gates did not respond, and the foundation continues to promote sterilizing vaccines, contraception, and abortions to the African people.

BMGF has given over $80 million to the abortion giant, Planned Parenthood, over the last decade.[30] The foundation has also donated over $40 million to a British abortion company called Marie Stopes. This company was kicked out of Kenya in 2018 for undermining the government and performing illegal abortions in the country.[31] Regardless of one's opinion about abortion, most people would agree that BMGF had no right to impose population control on countries that do not want it. Of course, Bill Gates thinks he knows best for everyone.

Sterilization is not the only byproduct of the Gates family vaccination program. The Bill and Melinda Gates Foundation committed $450 million to eradicate polio in India, taking control of India's National Technical Advisory Group on Immunization to implement its plan. Children were mandated to have several doses of polio vaccine before age five, with mass vaccination drives carried out 3 times per year in order to maintain herd immunity.[32] The result was an epidemic of non-polio acute flaccid paralysis (NPAFP), a sudden onset of paralysis or weakness in children under the age of 15. Approximately 490,000 additional children beyond expected rates were affected between 2010 and 2017[33] and the incidence of non-polio AFP was related to the frequency of polio vaccine administration.

The Program for Appropriate Technology in Health (PATH), funded by BMGF, carried out observational studies, administering the HPV vaccines Gardasil and Cervarix to thousands of girls between age 9 and 15 in India. Within months, girls started getting sick and several died. Reactions included epileptic seizures, severe stomachache, headaches, mood swings, early onset of menstruation, and heavy bleeding.

An investigation showed that there were significant problems with consent, which was sometimes given by school officials instead of parents. Illiterate parents consented with thumbprints on the forms. In one province, 3944 of the consent forms had thumb impressions and 5,454 had either signed or thumb impressions of guardians. For the most part, girls and their parents had no understanding of cervical cancer or the vaccine, or what they were agreeing to.[34]

In 2017 the Modi government threw Gates and his vaccine initiatives out of India.[35]

There are additional ethical issues concerning the Gates Foundation beyond the vaccination initiatives. It is not illegal for a 501(c)3 non-profit to invest in the stock market if there are no conflicts of interest. One of the foundation's more recent acquisitions was stock in a company called Schrodinger, which specializes in the development of new drugs.[36] BMGF has been involved in financial ventures with the company since 2010, and most recently an $85 million venture in 2019.[37] The company's website indicates that its promising products are COVID-19 treatments and vaccines.[38] The Gates Foundation currently is invested in Merck, GSK, Eli Lilly, Pfizer, Novartis and Sanofi.[39] The Gates Foundation has donated more than $300 million dollars to Inovio Pharmaceuticals, AstraZeneca, Moderna, and other drug companies to fund clinical trials for a COVID-19 vaccine.[40]

Not surprisingly, the Gates Foundation is staffed by several former drug company executives. Current CEO Penny Heaton was formerly employed by Merck and Novartis. Trevor Mundel, the foundation's president of global health, worked for Novartis and Pfizer. Tachi Tamanda, who held this position before him, was an executive with GlaxoSmithKline (GSK). Another GSK alumni, Kate James, is the foundation's communications director.[41]

Finally, Bill seems to think that the world is his to control. He stated during an interview on CBS This Morning on April 2nd, 2020: "Which activities, like schools, have such benefit and can be done in a way that the risk of transmission is very low?

And which activities, like mass gatherings, may be – in a certain sense – more optional? And so until you're widely vaccinated, those may not come back at all."[42] Thus Gates announced to the world the conditions under which he thinks humans can go back to normal living. And as of the time of the writing of this book, it appears that government and health officials are listening to him, since most of the world is still in some form of lockdown or restriction. After all, Bill Gates knows what is best for all of us.

Get to Know Robert Redfield

Robert Redfield, MD is the Director of the Centers for Disease Control and Prevention. The CDC's website describes Redfield as a "public health leader" who has been involved in both research and clinical care related to viral infections and infectious diseases for over 30 years.

His background sounds impressive. He spent 20 years working for the U.S. Army Medical Corps and was the founding director of the Department of Retroviral Research which was part of the U.S. Military's HIV program. He then co-founded the Institute of Human Virology at the University of Maryland and served as chief of Infectious Diseases and Vice Chair of Medicine at the University of Maryland School of Medicine.[43] Redfield certainly seems qualified for the job based on the information posted on the CDC's website.

But all is not as it seems. In 1994, Redfield was accused of overstating test results of an experimental AIDS vaccine made by MicroGeneSys. He was transferred from the lab he headed for six years to a much less prestigious position at Walter Reed Medical Center in Washington D.C. where he treated patients.[44] At the time Air Force officials insisted that this was not a demotion and that Redfield wanted to return to patient care.

There was, however, more to the story. When it became apparent that Redfield might be appointed to head the CDC, one of the Army whistleblowers, Air Force Lt. Col. Craig Hendrix, a doctor who heads the division of clinical pharmacology at Johns Hopkins University School of Medicine, decided to speak out about what really happened. Hendrix says that while Redfield was conducting HIV research for the army, "Either he was egregiously sloppy with data or it was fabricated. It was somewhere on that spectrum, both of which were serious and raised questions about his trustworthiness." Furthermore, according to Hendrix, two members of Redfield's team reported that they had tried to replicate his research findings but could not. Hendrix tried to replicate the findings and was unsuccessful, which is when he reported the problems with Redfield's data to his superiors.[45]

A meeting with Hendrix, Redfield, and others involved in the debacle was convened to discuss the issue, and according to Hendrix, Redfield admitted that he had overstated the results of his research. But Redfield made the same inaccurate representations after this meeting during a presentation at an international AIDS conference in Amsterdam in July 1992.

Just two months later, Congress appropriated $20 million for larger-scale testing of the MicroGeneSys vaccine. National Institutes of Health researchers disagreed with this decision, stating that scientists, not politicians, should be making decisions about which products would be studied in clinical trials.[46]

Redfield's continuous misrepresentations prompted Hendrix to write a formal letter accusing Redfield of scientific misconduct. An Air Force institutional review board recommended that the matter be investigated. But it wasn't. Instead, the Army stated that the data would be corrected, and Redfield was transferred.

Shocked that the matter had been unceremoniously dropped, in June 1994, Public Citizen wrote to Congressman Henry Waxman asking that his Subcommittee on Health and the Environment hold a hearing to investigate Redfield.

Here are excerpts from the letter, which provides more information about what happened:

"We are writing to request that your Subcommittee hold a hearing, as soon as possible, to investigate charges of grave impropriety committed by U.S. Department of Defense's AIDS researchers. We have obtained Internal memoranda, not previously made public, from the Department of Defense that allege a systematic pattern of data manipulation, inappropriate statistical analyses and misleading data presentation by Army researchers in an apparent attempt to promote the usefulness of the GP160 AIDS vaccine...

The Phase I and Phase II studies in which this alleged misconduct occurred were conducted by researchers at the Walter Reed Army Institute of Research (WRAIR), led by Lt. Col. Robert Redfield, M.D., Chief of the Department of Retroviral Research, and misleading results from these trials were reported in...the New England Journal of Medicine in June 1991, the Journal AIDS Research and Human Retroviruses in June 1992 and the annual International AIDS Conference in Amsterdam in July 1992. In addition, overstated conclusions have been presented on two occasions at hearings before your Subcommittee.

"Meeting on October 23, 1992 to discuss the allegations by two Air Force research physicians (see below) of scientific misconduct by Dr. Redfield, a subcommittee of the Institutional Review Committee at the Wilford Hall U.S. Air Force Medical Center, San Antonio, Texas reached the following conclusion:

"The committee agreed the information presented by Dr. Redfield seriously threatens his credibility as a researcher and has the potential to negatively impact AIDS research funding for military institutions as a whole. His allegedly unethical behavior creates false hope and could result in premature deployment of the vaccine...

"That meeting was called to review an October 21, 1992 memorandum...from Maj. Craig W. Hendrix, M.D., Director of the HIV Program in the Air Force, and Col. R. Neal Boswell, MD., Associate Chief of the

Division of Medicine in the Air Force, to Col. Donald Burke, M.D., Director of the Division of Retrovirology at WRAIR and Dr. Redfield's immediate supervisor. The memorandum decried 'The problem of misleading or, possibly, deceptive presentations by Dr. Redfleld, which overstate the GP160 [vaccine] Phase I data...' and recommended that the following action be taken:

"(1) publicly correct the record in a medium suitable for widespread dissemination to our civilian scientific colleagues;

"(2) censure Dr. Redfield for potential scientific misconduct which should at least include temporarily suspending his involvement on the current immunotherapy protocols; and

"(3) initiate an investigation by a fully independent outside Investigative body...to evaluate the facts of the case and recommend appropriate actions.

"Senior Department of Defense scientists have known of this misconduct since at least October 1992, and Dr. Redfield has acknowledged that his analyses were faulty on at least three occasions to internal Department of Defense audiences (the earliest admission was on August 28, 1992)..."[47]

At the time these events took place, there was tremendous pressure to address the AIDS crisis, and the National Institute of Allergy and Infectious Diseases (NIAID) Director Anthony Fauci incessantly promoted the idea that a vaccine was the answer. There would be tremendous financial rewards for the individual or individuals who developed an effective vaccine, and even the potential to win a Nobel Prize. Redfield had an incentive to fabricate data.

Even mainstream media outlet CNN covered this debacle recently, as well as the fact that health officials who worked with Redfield reported that he consistently demonstrated bad leadership, tended to be a bully, and usually put politics ahead of data.[48]

Redfield was not disciplined, and instead was given more opportunities in the healthcare field and placed in positions of leadership, all providing him with more opportunities to engage in the same misbehavior he had demonstrated repeatedly before.

Get to Know Deborah Birx

Deborah Birx M.D. is currently employed by the U.S. Department of State as the U.S. Global AIDS Coordinator and U.S. Special Representative for Global Health Diplomacy.[49]

Dr. Birx was recruited by Dr. Robert Redfield in 1988 to be his research assistant when he was conducting research on an AIDS vaccine while employed by the Army. At the time, co-workers described them as a tight unit, working together to test the MicroGeneSys vaccine.

When Redfield was accused of falsifying data concerning the efficacy of the vaccine, Birx was his chief defender, and questioned the motives of those who accused him.

As previously, Redfield was never charged with misconduct, nor was his assistant Birx, although the Army concluded he had violated Army code due to his relationship with Americans for Sound AIDS Policy (ASAP). The Army stated that ASAP had received scientific information from Walter Reed "to a degree that is inappropriate" and that ASAP seemed to be a vehicle "for marketing LTC Redfield's research." Redfield served on ASAP's "Science Advisory Board" along with Birx.

After both left the Army, Redfield continued to conduct AIDS research, and his Institute at the University of Maryland was the beneficiary of millions of dollars in grants approved by Birx, who ran the President's Emergency Plan for AIDS Relief (PEPFAR).[50] She was appointed by President Barack Obama.

Despite this questionable past, Birx was appointed as the Coronavirus Response Coordinator when the COVID-19 "crisis" began.

Birx sits on the board of the Institute for Health Metrics and Evaluation (IHME), which produced one of the wildly inaccurate models used to make decisions about COVID-19 and was an enthusiastic promoter of this model. IHME has received significant funding from the Bill and Melinda Gates Foundation.[51]

Birx also sits on the board of the Global Fund, also funded by the bill and Melinda Gates Foundation. The Foundation gave the Global Fund $750 million in 2012.[52]

According to her biography posted on a government site, Birx "helped lead one of the most influential HIV vaccine trials in history (known as RV 144 or the Thai trial), which provided the first supporting evidence of any vaccine's potential effectiveness in preventing HIV infection."[53] Decades later, and after spending billions of dollars, there still is no vaccine for HIV.

Get to Know Anthony Fauci

On the surface it appears that Dr. Anthony "Tony" Fauci has an accomplished track record. He is the head of the National Institute of Allergy and Infectious Diseases (NIAID), which is a division of the National Institutes of Health (NIH). He was appointed in 1984 under the Reagan administration, and has served in this position under every president since. He was appointed as a member of the Coronavirus Task Force by President Trump.

His resume is impressive. He graduated first in his class from Cornell Medical School. He has received many awards and accolades during his career, including The 2016 International AIDS Society President's Award (07/19/2016). He was named a 2019 Distinguished Fellow of the American Association of Immunologists (03/22/2019), and was honored for 35 Years of Leadership in HIV Policy and Research by AIDS United's Public Policy Council (10/02/2019).

He seems knowledgeable and to be an expert. But not all is as it seems. Fauci, like many non-elected officials who hold their positions for a significantly long period of time, has made friends with many important politicians, drug company executives, and influencers like Bill Gates. His decisions indicate an unwillingness to admit that he is wrong, as well as conflicts of interest. Yet he remains a powerful person, more so than most elected officials.

An important question to ask is, "What has he done with his power?"

HIV Hero or Villain?

Fauci began his career with the National Institutes of Health in 1968 and by 1977 he had risen to deputy clinical director of the NIAID. He had never practiced medicine, was not known as a great scientist or researcher, and had never managed a large agency. He did, however, have a lot of ambition and the HIV/AIDS crisis presented a great opportunity for him.

In 1985 NIAID received a small amount of funding for HIV/AIDS, but over time this grew to billions of dollars of funding, and was a significant percentage of the total budget for the NIH.[54] Fauci's hypothesis was that all AIDS cases were caused by HIV, and that his agency should focus on assisting in the development of a vaccine.

Unfortunately for him, inconvenient data began to surface which showed that not all people with AIDS were infected with HIV. When confronted with the fact that 4621 clinically diagnosed AIDS patients were HIV negative, Fauci, in conjunction with the CDC, renamed these cases as "idiopathic CD-4 lymphcytopenia (ICL)."[55]

Fauci also chose to ignore the fact that HIV/AIDS did not share common characteristics of infectious diseases, which do not discriminate based on sex. This is true of all known infectious diseases including flu, polio, hepatitis, tuberculosis, and pneumonia. AIDS, on the other hand, developed in a few high-risk groups such as intravenous drug users and gay males using recreational drugs, occurred in ten times as many men as women and preferred gay over straight men.

Infectious diseases spread exponentially, but AIDS did not follow this trajectory. It steadily increased from a few dozen cases in 1981 to over 80,000 cases in 1994. It did not explode and become widely spread, nor did it decline which is typical as a population gains herd immunity.[56]

In other words, Fauci ignored a growing body of evidence showing that his hypothesis was incorrect, but would not change his mind, his focus, or the activities of his agency. But that's not all.

AZT stands for azidothymidine, a drug originally developed as a treatment for cancer. The drug was not only ineffective for treating cancer, but mice treated with it died of extreme toxicity.[57] It was never patented until the company that owned it, Burroughs-Wellcome, proposed using it to treat AIDS patients based on its ability to prevent HIV from multiplying in a test tube.

Executives from Burroughs lobbied the FDA to begin clinical trials immediately. While clinical trials are supposed to be double-blind and placebo controlled, everyone soon knew which patients were taking AZT due to the horrific side effects of the drug. But the FDA approved AZT for the treatment of AIDS patients anyway, in part due to public pressure to find a cure.

Even with the known side effects of the drug, NIAID, under Fauci's direction, decided to conduct a clinical trial of AZT on pregnant mothers with HIV who were also addicted to intravenous drugs. The trial was ended early when Fauci and his collaborators announced that they had reduced HIV transmission by two thirds – from 25% to 8% with AZT treatment.[58] The data showed that 13 out of 180 babies born to AZT-treated mothers were HIV-positive as compared to 40 out of 184 babies born to mothers given a placebo.

Fauci seemed to ignore the fact that most babies were not born HIV positive in the placebo group, and to save 27 babies, 180 mothers and 153 of their unborn babies were given a toxic drug with significant side effects. The early termination of the clinical trial meant that AZT treatment to HIV positive pregnant mothers would become a standard of care without any long-term follow-up on the effect on the mothers or their babies. By this time Fauci knew that HIV was not the cause of AIDS, which made the promotion of this treatment appear even more careless.

In 2008, after squandering billions of dollars on HIV vaccine research, and authorizing a questionable clinical trial on pregnant women, Fauci admitted that little was known about HIV. He said that out of the tens of millions of people who had been infected with HIV, there was not one documented case of a person who was infected and cleared the virus. This, according to Fauci, meant that "...we don't even know if the body is capable of eliciting a protective immune response." He also acknowledged that there were many people referred to as "long-term non-progressors" who are somehow able to live for a long time with the virus.[59]

Despite his incompetence, Fauci continued to head NIAID. Unfortunately for the American people he was considered the best person to lead the response team when the COVID-19 debacle began.

Discerning a Pattern

Fauci, Birx, and Redfield share several common traits. Once they latch onto an idea, nothing will dissuade them from continuing to pursue it, including evidence that the idea is no longer viable. All three appear to be ethically challenged. And they all seem to end up in the same place at the same time – again, and again, and again.

ENDNOTES

1. Ethiopia: Donor Aid Supports Repression. Human Rights Watch. October 19, 2010. https://www.hrw.org/news/2010/10/19/ethiopia-donor-aid-supports-repression#. Accessed August 28, 2020.

2. 2.5 Million Amharas missing confirmed in parliament part 1 (video). YouTube. Published October 20, 2012. https://www.youtube.com/watch?v=ndSzSPskwAw&t=34s&app=desktop. Accessed August 28, 2020.

3. The case of the 2.5 million missing Amharas in Ethiopia: Timing and ensuring the fairness and transparency of the upcoming census. Reddit. https://www.reddit.com/r/Africa/comments/as4jmy/the_case_of_the_25_million_missing_amharas_in/. Accessed August 28, 2020.

4. Bartels SA, Greenough PG, Tamar M, VanRooyan MJ. Investigation of a cholera outbreak in Ethiopia's Oromiya Region. *Disaster Med Public Health Prep.* 2010;4(4):312-317.

5. Donald McNeil. Candidate to Lead the W.H.O. Accused of Covering Up Epidemics. *New York Times.* May 13, 2017. https://www.nytimes.com/2017/05/13/health/candidate-who-director-general-ethiopia-cholera-outbreaks.html. Accessed August 28, 2020.

6. Audit Reports and Diagnostic Review issued by the Global Fund's Office of the Inspector General on 20 April 2012. https://www.theglobalfund.org/media/2686/oig_gfoig10014auditethiopia_report_en.pdf. Accessed September 1, 2020.

7. Tedros Adhanon Incompetency, Politics and Immorality: Full Documentary. YouTube. https://www.youtube.com/watch?v=_aEOUXrilhw. April 22, 2017. Accessed September 1, 2020.

8. Ibid.

9. Robert Mugabe, Zimbabwe's strongman ex-president, dies aged 95. *BBC News.* September 6, 2019. https://www.bbc.com/news/world-africa-49604152. Accessed September 1, 2020.

10. Bill Gates Biography. Biography. Updated April 8, 2020. https://www.biography.com/business-figure/bill-gates. Accessed September 1, 2020.

11. Associated Press. Mary Gates, 64; Helped Her Son Start Microsoft. *New York Times.* June 11, 1994. https://www.nytimes.com/1994/06/11/obituaries/mary-gates-64-helped-her-son-start-microsoft.html. Accessed September 1, 2020.

12. Dominic Otieno. Bill Gates Sr. Information Cradle. https://informationcradle.com/bill-gates-sr/. Updated June 26, 2020. Accessed September 1, 2020.

13. Hitler, The Ku Klux Klan, and Margaret Sanger. https://www.courierherald.com/letters/hitler-the-ku-klux-klan-and-margaret-sanger/ Updated June 26, 2020. Accessed September 1, 2020.

14. Willis Krumholz. Yes, Planned Parenthood Targets And Hurts Poor Black Women. The Federalist. https://thefederalist.com/2016/02/18/yes-planned-parenthood-targets-and-hurts-poor-black-women/. February 18, 2016. Accessed September 1, 2020.

15. Our History. The Rockefeller Foundation. https://www.rockefellerfoundation.org/about-us/our-history/#:~:text=Leading%20philanthropists%2C%20including%20Richard%20Chandler,in%20health%2C%20education%2C%20and%20economic. Accessed September 2, 2020.

16. Rockefeller Foundation. Influence Watch. https://www.influencewatch.org/non-profit/rockefeller-foundation/#population-control. Accessed September 2, 2020.

17. Health and Well-Being. Science, Medical Education and Public Health. The Rockefeller Foundation Centennial Series. https://www.rockefel-lerfoundation.org/wp-content/uploads/Health-Well-being.pdf. Accessed September 2, 2020.

18. Ibid.

19. Scott Ard, Wylie Wong, John Borland. Judge rules Microsoft violat-ed antitrust laws. Cnet. https://www.cnet.com/news/judge-rules-mic-rosoft-violated-antitrust-laws/. January 2, 2002. Accessed September 1, 2020.

20. John Fontana. Gates legacy filled with good, bad and ugly. Net Work World. https://www.networkworld.com/article/2280597/gates-leg-acy-filled-with-good--bad-and-ugly.html. June 10, 2008. Accessed September 1, 2020.

21. Martin Bürger. Bill Gates: Life won't go back to 'normal' un-til population 'widely vaccinated'. Life Site News. https://www.lifesitenews.com/news/bill-gates-life-wont-go-back-to-normal-until-population-widely-vaccinated. April 6, 2020. Accessed September 1, 2020.

22. Awarded Grands. Bill & Melinda Gates Foundation. https://www.gatesfoundation.org/how-we-work/quick-links/grants-database#q/k=Population%20Council. Accessed September 1, 2020.

23. Population Policy and Demographic Analysis. Population Council. https://www.popcouncil.org/research/population-policy-and-demo-graphic-analysis. Accessed September 1, 2020.

24. Emily Flitter, James B. Stewart. Bill Gates Met With Jeffrey Epstein Many Times, Despite His Past. *New York Times.* https://www.nytimes.com/2019/10/12/business/jeffrey-epstein-bill-gates.html. Updated November 26, 2019. Accessed September 1, 2020.

25. The Pastors Chronicles. UN Vaccines Sterilize 500,000 Women in Kenya. The Pastors Chronicles. https://tpchronicles.com/

un-vaccines-sterilize/. December 8, 2018. Accessed September 1, 2020.

26. Jacob Puliyel. Ethical questions surround vaccine to reduce fertility. *Sunday Guardian Live.* https://www.sundayguardianlive.com/news/ethical-questions-surround-vaccine-reduce-fertility. May 26, 2018. Accessed September 1, 2020.

27. Oller J, Shaw CA, Tomljenovic L, Karanja SK, et al. HCG Found in WHO Tetanus Vaccine in Kenya Raises Concern in the Developing World. Open Access Library Journal. *ResearchGate.* 2017;4(10):1-30.

28. Sean Adl-Tabatabai. Renowned Scientist: Bill Gates Is Funding 'Population Control' in Africa. News Punch. https://newspunch.com/scientist-bill-gates-funding-population-control-africa/. January 26, 2019. Accessed September 1, 2020.

29. Obianuju Ekeocha. An African Woman's Open Letter to Melinda Gates. Pontifical Council for the Laity. http://www.laici.va/content/laici/en/sezioni/donna/notizie/an-african-woman-s-open-letter-to-melinda-gates.html. Accessed September 1, 2020.

30. Joseph Vazquez. Bill Gates Sends Over $11 Million to Pro-Abortion Group Pushing Population Control Worldwide. LifeNews.Com. https://www.lifenews.com/2020/03/20/bill-gates-sends-over-11-million-to-pro-abortion-group-pushing-population-control-worldwide/. March 20, 2020. Accessed September 1, 2020.

31. Fr. Shenan J. Boquet. Kenya…and the Marie Stopes and Gates Foundation Onslaught. Human Life International. https://www.hli.org/2018/11/kenya-and-the-gates-foundation/. November 26, 2018. Accessed September 1, 2020.

32. Centers for Disease Control and Prevention. Progress Toward Poliomyelitis Eradication – India January 2009-October 2010. *Morbidity and Mortality Weekly Report.* 2010 Dec;59(48):1581-1585.

33. Dhiman R, Prakash SC, Screenivas V, Puliyel J. Correlation between Non-Polio Acute Flaccid Paralysis Rates with Pulse Polio Frequency in India. *Int J Environ Res Public Health* 2018;15(8):1755. doi: 10.3390/ijerph15081755.

34. KP Narayana Kumar. Controversial vaccine studies: Why is Bill and Melinda Gates Foundation under fire from critics in India? *The Economic Times.* https://economictimes.indiatimes.com/industry/healthcare/biotech/healthcare/controversial-vaccine-studies-why-is-bill-melinda-gates-foundation-under-fire-from-critics-in-india/articleshow/41280050.cms?from=mdr. August 31, 2014. Accessed September 1, 2020.

35. BS Web Team. Modi govt cuts ties with Bill and Melinda Gates Foundation on immunization. *Business Standard.* https://www.business-standard.com/article/economy-policy/modi-govt-cuts-ties-with-bill-and-melinda-gates-foundation-on-immunisation-117020800294_1.html#:~:text=Modi%20govt%20cuts%20ties%20with%20Bill%20and%20Melinda%20Gates%20Foundation%20on%20immunisation,-BS%20Web%20Team&text=All%20financial%20ties%20between%20the,the%20Economic%20Times%20on%20Wednesday. February 8, 2017. Accessed September 1, 2020.

36. Bill Gates' Portfolio. GuruFocus. https://www.gurufocus.com/guru/bill+gates/current-portfolio/portfolio. Accessed September 1, 2020.

37. Stephanie Simon. Schrödinger Closes $85 Million Financing to Advance Computational Platform and Expand Drug Discovery Pipeline. BusinessWire. https://www.businesswire.com/news/home/20190104005471/en/Schrödinger-Closes-85-Million-Financing-Advance-Computational. January 4, 2019. Accessed September 1, 2020.

38. https://www.schrodinger.com/blog/google-covid-19 accessed 9.2.2020

39. Tim Schwab. Bill Gates's Charity Paradox. *The Nation.* https://www.
     thenation.com/article/society/bill-gates-foundation-philanthropy/.
     March 17, 2020. Accessed September 1, 2020.

40. Ciara Linnane. Inovio stock rallies after company gets Gates
     Foundation grant to test device for coronavirus vaccine.
     MarketWatch. https://www.marketwatch.com/story/inovio-stock-
     rallies-after-company-gets-gates-foundation-grant-to-test-device-
     for-coronavirus-vaccine-2020-03-12. March 14, 2020. Accessed
     September 1, 2020.

41. Michele Greenstein, Jeremy Loffredo. Why the Bill Gates global
     health empire promises more empire and less public health. *The
     Grayzone.* https://thegrayzone.com/2020/07/08/bill-gates-global-
     health-policy/. July 8, 2020. Accessed September 1, 2020.

42. Martin Bürger. Bill Gates: Life won't go back to 'normal' un-
     til population 'widely vaccinated'. Life Site News. https://www.
     lifesitenews.com/news/bill-gates-life-wont-go-back-to-normal-until-
     population-widely-vaccinated. April 6, 2020. Accessed September 1,
     2020.

43. Centers for Disease Control and Prevention. Director: Robert R.
     Redfield, MD. https://www.cdc.gov/about/leadership/director.htm.
     April 16, 2018. Accessed September 1, 2020.

44. Lyn Bixby. Army's Top AIDS Researcher Transferred Amid
     Controversy. *Hartford Courant.* June 30, 1994.

45. Taylor M. "Research Misconduct Allegations Shadow New CDC
     Head." *Kaiser Health News* March 21 2020

46. Bixby L. "Army's Top AIDS Researcher Transferred Amid
     Controversy." *Hartford Courant.* https://www.courant.com/news/
     connecticut/hc-xpm-1994-06-30-9406300185-story.html. June 30
     1994. Accessed September 1, 2020.

47. *Kaiser Health News.* https://khn.org/wp-content/uploads/ sites/2/2018/03/940607plswtowaxman.pdf. June 7, 1004. Accessed September 1, 2020.

48. Kristen Holmes, Nick Valencia, Curt Devine. CDC woes bring Director Redfield's troubled past as an AIDS researcher to light. *CNN.* https://www.cnn.com/2020/06/04/politics/cdc-red-field-aids-walter-reed-army-investigation/index.html. June 5, 2020. Accessed September 1, 2020.

49. Deborah L. Birx, M.D. U.S. Department of State. https://www. state.gov/biographies/deborah-l-birx-md/. April 2, 2014. Accessed September 1, 2020.

50. Kristen Holmes, Nick Valencia, Curt Devine. CDC woes bring Director Redfield's troubled past as an AIDS researcher to light. *CNN.* https://www.cnn.com/2020/06/04/politics/cdc-red-field-aids-walter-reed-army-investigation/index.html. June 5, 2020. Accessed September 1, 2020.

51. Patrick Howley . Gates Foundation Funded BOTH Imperial College and EHME, Failed Model Makers. *National File.* https://nationalfile. com/gates-foundation-funded-both-imperial-college-and-ihme-failed-model-makers/. May 16, 2020. Accessed September 1, 2020.

52. Bill & Melinda Gates Foundation. Foundation Commits $750 Million to Global Fund | Bill & Melinda Gates Foundation. https:// www.gatesfoundation.org/Media-Center/Press-Releases/2012/01/ Foundation-Commits-$750-Million-to-Global-Fund. Accessed September 1, 2020.

53. Deborah L. Birx, M.D. U.S. Department of State. https://www. state.gov/biographies/deborah-l-birx-md/. April 2, 2014. Accessed September 1, 2020.

54. The AIDS Research Program of the National Institutes of Health. Supporting the NIH AIDS Research Program. https://www.ncbi.nlm. nih.gov/books/NBK234085/. Accessed September 1, 2020.

55. Malaspina A, Moir S, Chaitt DG et al. "Idiopathic CD4+ T lympho-cytopenia is associated with increases in immature/transitional B cells and seru levels of Il-7." *Blood.* 2007 Mar 1;109(5):2086-2088. doi: 10.1182/blood-2006-06-031385.

56. Bergman DJ, Langmuir AD. Farr's Law Applied to AIDS Projections. *JAMA.* 1990;263(11):1522-1525. doi:10.1001/jama.1990.03440110088033.

57. Lauritsen J. *Poison by Prescription: The AZT Story.* New York, NY: Asklepios Press; 1990.

58. Connor EM, Sperling RS, Gelber R et al. Reduction of mater-nal-infant transmission of human immunodeficiency virus type 1 with zidovudine treatment. Pediatric AIDS Clinical Trials Group Protocol 076 Study Group. *N Engl J Med.* 1994;331(18):1173-1180. doi:10.1056/NEJM199411033311801.

59. Nikhil Swaminathan. NIH Official: HIV Vaccine Research "Swimming in the Dark". *Scientific American.* https://www.scientificamerican.com/article/nih-official-fauci-hiv-vaccine/. July 28, 2008. Accessed September 1, 2020.

THE GLOBAL HEALTH AND VACCINE CABAL

DURING THE LAST several years, a few individuals, foundations, and companies have partnered with international organizations and together, have formed even more international organizations. These organizations share many of the same objectives, receive funding from the same entities, and share many of the same individual and corporate members. The fact that there are so many helps to create the illusion that a "global consensus" has developed concerning healthcare policies, practices, and vaccines in particular. The reality is that a rather small number of individuals and drug companies have managed to gain control over healthcare, which is an important part of the lives of the entire world population. And while these people have managed to disguise their interrelationships, they are quite clear about their ultimate objective: mandatory vaccines for all world citizens. They think that vaccines should be required for all as a condition for freedoms we currently take for granted.

The freedom to move about, travel, work, shop, and congregate with others.

Some of the organizations, such as the Rockefeller Foundation, were formed a long time ago, and have a long history of engaging in questionable activities. Others were founded more recently, and new ones are regularly started by the same people and groups. A few have engaged in activities that seem to indicate that the COVID-19 pandemic was planned, or at least contemplated for a long period of time.

The Global Health and Vaccine Cabal is a confusing matrix of organizations. We struggled with how to present this very important information in a way that was coherent and decided that it was best to just list them, and allow you to see how the same people appear again, and again, along with the same primary funding organization – **the Bill and Melinda Gates Foundation**.

The Fascinating History of the Rockefeller Foundation

At first glance, the Rockefeller Foundation appears to be a benevolent organization created by one of America's wealthiest families. Unfortunately, since its founding in 1913, the foundation has engaged in and funded a series of projects that indicate a far more sinister agenda.

John D. Rockefeller established his namesake foundation "to promote the well-being of mankind throughout the world." The foundation joined the American movement of "social philanthropy," launched by Andrew Carnegie to use wealth to support "...social improvement...order, productivity, and secular advancement." In addition to its focus on public health, the foundation engaged in geopolitical activities that included influencing foreign governments and how they responded to their citizens, expanding consumer markets, and promoting the internationalization of science and culture.[1]

The foundation's influence has resulted in several practices that have been described as "marked asymmetries in political medical power." These include top-down agenda setting, using financial incentive to influence outcomes, working toward international consensus rather than localized decision-making, and supporting social medicine practices that are integrated with political ambitions and outcomes.[2]

While some of the practices described above might be cause for concern for some, other foundation projects are outright alarming. In the early 1900s, the Rockefeller Foundation, along with many other wealthy individuals and groups, funded research on eugenics, a practice originating with Frances Galton, Darwin's cousin. In 1904 Galton published *Eugenics: Its Definition, Scope and Aims,* which outlined how the U.S. could improve its population through selection of people who had certain traits such as "health, energy, ability, manliness, and courteous disposition." He defined eugenics as "the science which deals with all influences that inborn qualities of a race." He stated in his article that "most savage races" disappear when brought into contact with civilization, but "some, like the negro, do not." This was an argument, he said, for bringing "as many influences as can be reasonably employed, to cause the useful classes in the community to contribute *more* than their proportion to the next generation."[3]

John D. Rockefeller was interested in eugenics, particularly population control. He joined the American Eugenics Society, and served as a trustee of the Bureau of Social Hygiene.[4] In correspondence with Charles Davenport, Director of the Eugenics Records, Rockefeller suggested that certain women should be incarcerated for longer than their actual sentence so that they "would…be kept from perpetuating their kind… until after the period of child-bearing had passed."[5]

From the beginning, Rockefeller wanted to promote eugenics worldwide, so it was not surprising that the foundation decided to support eugenics research in Germany. The Germans were interested in eugenics before Hitler took office, and the largest contributor to eugenics programs in Germany was the Rockefeller Foundation. By 1926 the foundation had donated $410,000 (the equivalent of $4 million today) – to hundreds of German eugenics researchers.[6]

One of the foundation's larger grants, $250,000 was given to the German Institute for Psychiatric Research, which studied issues such as the relationship between race and mental illness. Researchers "concentrated on locating the genetic and neurological basis of traits such as criminal propensity and mental disease" and the relationship between mental retardation and race.[7] Funding continued for at least two years after Hitler and the Nazis rose to power. Ernst Rudin, the head of the Institute, provided considerable support for Hitler's racist policies.

In 1932, Eugen Fischer, head of the Kaiser Wilhelm Institute for Anthropology, Human Heredity, and Eugenics (KWIA) and architect of the German Mapping Project, sought funding from the Rockefeller Foundation to study, among other things, the differences between German Jews and pure Germans and to determine the effects of interbreeding.[8] Additionally, KWIA conducted research on twins. One of Fischer's projects involved injecting twins with substances such as lead and mercury to determine the impact on the children and future generations. It was determined that identical twins reacted more similarly than fraternal twins.[9]

The Rockefeller Foundation (RF) started funding childhood vaccination programs in 1984, after sponsoring an international meeting at the Rockefeller Foundation conference center in Bellagio Italy. The World Health Organization had established the Expanded Program on Immunization (EPI) in 1980, and after the Bellagio meeting, hundreds of millions of dollars were donated to EPI.[10]

The Children's Vaccine Initiative (CVI) was started in 1990 and was a collaboration between the United Nations Children's Fund (UNICEF), the World Health Organization (WHO), the World Bank and the Rockefeller Foundation. RF donated $2.5 million to CVI.[11]

The connection between the Rockefeller Foundation and COVID-19 may have begun in 2010, when the foundation published a report in partnership with the Global Business Network (GBN) titled "Scenarios for the Future of Technology and International Development."[12] GBN chairman and co-author of the report Peter Schwartz described the project as "ambitious" and said that "The [Rockefeller] Foundation has stretched its thinking far beyond theoretical models of technology innovation and diffusion in order to imagine how technology could actually change the lives of people from many walks of life." Judith Rodin, president of the foundation and co-author of the report said, "I hope this publication makes clear exactly why my colleagues and I are so excited about the promise of using scenario planning to develop robust strategies and offer a refreshing viewpoint on the possibilities that lie ahead."[13]

The collaboration used "scenario planning" to look at possible responses to hypothetical situations, including pandemics. The authors note that the scenarios described in the report are designed to be "plausible" and to explore what the future might look like. The points covered include:

- A timeline of headlines and events as they unfold
- A description of the technology that might be used
- Discussion of the role of philanthropy and philanthropic organizations and how they might contribute
- A "day in the life" of a person who is living through the scenario

The first scenario is titled "LOCK STEP; a world of tighter top-down government control and more authoritarian leadership with innovation and growing citizen pushback." It describes a pandemic that takes place in 2012 that is virulent and deadly. Even well prepared countries are quickly overwhelmed and 20%

of the population becomes infected. The report includes these statements:

> "The pandemic also had a deadly effect on economies: international mobility of both people and goods screeched to a halt, debilitating industries like tourism and breaking global supply chains. Even locally, normally bustling shops and office buildings sat empty for months, devoid of both employees and customers."

> "The United States' initial policy of "strongly discouraging" citizens from flying proved deadly in its leniency, accelerating the spread of the virus not just within the U.S. but across borders. However, a few countries did fare better—China in particular. The Chinese government's quick imposition and enforcement of mandatory quarantine for all citizens, as well as its instant and near-hermetic sealing off of all borders, saved millions of lives, stopping the spread of the virus far earlier than in other countries and enabling a swifter postpandemic recovery."

> "During the pandemic, national leaders around the world flexed their authority and imposed airtight rules and restrictions, from the mandatory wearing of face masks to body-temperature checks at the entries to communal spaces like train stations and supermarkets. Even after the pandemic faded, this more authoritarian control and oversight of citizens and their activities

stuck and even intensified. In order to protect themselves from the spread of increasingly global problems—from pandemics and transnational terrorism to environmental crises and rising poverty—leaders around the world took a firmer grip on power."

"At first, the notion of a more controlled world gained wide acceptance and approval. Citizens willingly gave up some of their sovereignty—and their privacy—to more paternalistic state in exchange for greater safety and stability. Citizens were more tolerant, and even eager, for top-down direction and oversight, and national leaders had more latitude to impose order in the ways they saw fit. In developed countries, this heightened oversight took many forms: biometric IDs for all citizens, for example, and tighter regulation of key industries whose stability was deemed vital to national interests. In many developed countries, enforced cooperation with a suite of new regulations and agreements slowly but steadily restored both order and, importantly, economic growth."

"By 2025, people seemed to be growing weary of so much top-down control and letting leaders and authorities make choices for them. Wherever national interests clashed with individual interests, there was conflict. Sporadic pushback became increasingly organized and coordinated, as disaffected youth and people who had seen their status and opportunities slip away—largely in developing countries— incited civil unrest. In 2026, protestors in Nigeria brought down the government, fed up with the entrenched cronyism and corruption. Even those who liked the greater stability and predictability of this world began to grow uncomfortable and constrained by so many tight rules and by the strictness of national boundaries. The feeling lingered that sooner or later, something would inevitably upset the neat order that the world's governments had worked so hard to establish."

Technology in Lockstep is largely driven by government and is focused on issues of national security, health, and safety:

> Scanners using advanced functional magnetic resonance imaging (fMRI) technology become the norm at airports and other public areas to detect abnormal behavior that may indicate "antisocial intent."

> New diagnostics are developed to detect communicable diseases. The application of health screening also changes; screening becomes a prerequisite for release from a hospital or prison, successfully slowing the spread of many diseases.

The Role of Philanthropy in Lockstep:

> Philanthropy grantee and civil society relationships will be strongly moderated by government,

and some foundations might choose to align themselves more closely with national official development assistance (ODA) strategies and government objectives.

Philanthropic organizations interested in promoting universal rights and freedoms will get blocked at many nations' borders. Developing smart, flexible, and wide-ranging relationships in this world will be key; some philanthropies may choose to work only in places where their skills and services don't meet resistance.[14]

Many people might conclude that either the authors of this document had incredible psychic powers or that the document is being used worldwide today as a handbook for decision-making.

This document is not posted on the Rockefeller Foundation site, and the work is no longer referenced in its annual reports. There is also no information on the website about eugenics and collaboration with the Nazis. This seems to indicate that foundation personnel know that some of the organization's activities and ideas might not be viewed favorably by at least some members of the public.

We were able to find online various interviews in which co-author Peter Schwartz discussed the merits of scenario planning; news releases about this report; and it was posted on Goodreads on May 17 2020.[15] We feel certain that the document is valid.

The Bill and Melinda Gates Foundation's Byzantine Vaccine Empire

Bill and Melinda Gates and their namesake foundation are enthusiastic advocates for and promoters of vaccines. In his 2011 annual letter, Bill Gates wrote, "In the same way that during my Microsoft career I talked about the magic of software, I now spend my time talking about the magic of vaccines." One goal of the Bill and Melinda Gates Foundation (**BMGF**) is to accelerate "the development and commercialization of novel vaccines and the sustainable manufacture of existing vaccines." To that end, the organization invests "in expertise and platform technologies that help us make vaccines faster, better, and cheaper."

Indeed, the BMGF has invested billions of dollars in vaccine research and the promotion of vaccines. The Foundation, along with several other partners, including the Rockefeller Foundation, WHO, World Bank and UNICEF, launched the **Global Alliance for Vaccines and Immunisation (GAVI)** in 2000.[16] The foundation provided $750 million to start, and to date has provided a total of $4.1 billion to GAVI.[17] The Rockefeller Foundation has continued to provide funding to GAVI,[18] along with the U.S. government, which by 2019 had donated $2.5 billion out of the almost $19 billion GAVI has received.[19]

GAVI describes itself as a public/private partnership, and states that it "...represents the sum of its partners individual strengths, from WHO's scientific expertise and UNICEF's procurement system to the financial know-how of the world bank and the **market knowledge of the vaccine industry**." (emphasis ours).[20]

The International Federation of Pharmaceutical Manufacturers and Associates (IFPMA) has one seat on the Gavi Board. IFPMA represents over 55 members including vaccine makers Johnson and Johnson, GlaxoSmithKline, Merck & Co., Novartis, Sanofi Pasteur, and the vaccines division of Sanofi-Aventis and Pfizer.[21] GAVI lists as its partners WHO, World Bank, UNICEF, governments of donor countries, and governments of both developing and industrialized countries that are recipients of vaccines.[22]

In 2010 while at the World Economic Forum Annual Meeting, Bill and Melinda Gates were joined by Julian Mob-Levyt, CEO of the GAVI Alliance and together announced a Foundation donation of $10 billion for vaccine research. Bill said, "We must make this the decade of vaccines," and Melinda said "Vaccines are a miracle…"[23] Governments enthusiastically jumped on board, and 194 member states endorsed the **Global Vaccine Action Plan (GVAP)** at the 65th World Health Assembly.[24]

The Global Fund to Fight Aids, Tuberculosis and Malaria was formed in 2002 to raise money and invest in solutions for infectious disease and is based in Geneva Switzerland. Several governments including the U.S. contribute, and the largest private sector donor is the **BMGF**, which has donated $2.24 billion dollars to date.[25] Other investors include The Rockefeller Foundation and Takeda Pharmaceutical.[26]

In 2019, the Global Fund's Sixth Replenishment conference in Lyon France raised over one billion dollars, of which $760 million was donated by the **BMGF**. It was announced at this meeting that the money would be used to develop a tuberculosis vaccine and a matchstick sized implant that can be inserted under the skin that protects an individual from the HIV virus for one year.[27]

Deborah Birx sits on the board of the Global Fund,[28] and as previously noted was appointed Coronavirus Response Coordinator for the White House when the pandemic began.

In 2012, The World Health Organization, UNICEF, The National Institute of Allergy and Infectious Diseases (NIAID, with Dr. Fauci as Director) and the **BMGF** announced a collaboration with 194 member states called **the Global Vaccine Action Plan (GVAP).** Board members included:

- Dr. Anthony Fauci, Director NIAID
- Dr. Margaret Chan, Director General of WHO, who was found culpable in promoting a fake pandemic in 2009-2010
- Dr. Tachi Yamada, President of Global Health at the Bill and Melinda Gates Foundation

Steering Committee Members included:

- Dr. Nicole Bates, Senior Program Officer, Global Health Policy & Advocacy **BMGF** Helen Evans, Acting CEO GAVI Alliance
- Dr. Lee Hall, Chief, Parasitology and International Programs Branch, Division of Microbiology and Infectious Diseases, NIAID[29]

The Strategic Advisory Group of Experts on Immunization (SAGE) Decade of Vaccines Working Group was formed by the WHO to review the Decade of Vaccines Global Action Plan (GVAP) and assess progress. Secretariat members of this group include:

- Magdalena Robert from the **BMGF**
- Hope Johnson from the GAVI Alliance
- Lee Hall from the NIAID
- Robin Nandy from UNICEF
- Joachin Hombach from WHO[30]

In October 2014, Frances Collins, director of the National Institutes of Health, announced a "New Phase of Cooperation Between NIH and the BMGF." The purpose of this partnership, which was the outcome of a planning session between the two organizations, was to promote, among other things, vaccines.[31]

The NIH and BMGF began their working relationship when the **Grand Challenges in Global Health** was instituted, which is funded, in part, by the **BMGF**,[32] and the Foundation for the National Institutes of Health.[33] Grand Challenges began by focusing on 14 major challenges that could, if solved, improve health in the developing world, six of which are related to vaccines.[34] By 2014, the Foundation had given Grand Challenges one billion dollars,[35] and in 2019, nine out of 25 awarded grants were for vaccine-related projects.[36]

Members of the scientific board of Grand Challenges at the time of its formation included

- Richard Klausner, **BMGF**
- Sir Roy Anderson from the Imperial College of London – has served as an advisor to WHO, **BMGF**, and as Governor of the Wellcome Trust (supports development and distribution of vaccines).
- Christine M. Debouck, GlaxoSmithKline Pharmaceuticals
- Anthony S. Fauci, National Institute of Allergy and Infectious Diseases, NIH William H. Foege, **BMGF**
- Yiming Shao, Chinese Center for Disease Control and Prevention

The Foundation for the National Institutes of Health (FNIH) creates public-private partnerships to fight against disease in the U.S. and throughout the world, funding research, sponsoring educational events and symposia, and training researchers.[37] Major donors include numerous drug companies and the **BMGF**. In 2017 tens of millions of dollars were donated by drug companies and the **BMGF**.[38]

In April 2020, the NIH announced a public-private partnership called Accelerating COVID-19 Therapeutic Interventions and Vaccines. The partnership included 16 drug companies and is to be orchestrated and directed by the FNIH.[39]

The National Institute of Allergy and Infectious Disease is a branch of the National Institutes of Health, and Anthony Fauci has been director since 1984. **The Vaccine Research Center** conducts research and assists in the development of vaccines and has received funding from the **BMGF**.[40]

In 2015, world leaders agreed to a set **of Sustainable Development Goals, or SDGs,** concerning vaccines. In an article published in the *Lancet*, a group of enthusiastic vaccine advocates write, "the global health community must continue working to provide all vaccines recommended by WHO to each and every child by ramping up efforts to extend full immunisation to the about 19 million children who are still not fully protected against a core set of vaccine-preventable diseases."

They also wrote, "Although governments are the main providers of immunisation, the GVAP's success depends upon many stakeholders—families, communities, health professionals, civil society, development partners, global agencies, manufacturers, media, and the private sector." The article was authored by Margaret Chan, then Director-General of the WHO; Christopher Elias with the **BMGF**; Anthony Fauci, Director of NAIAD; Anthony Lake of the UN Children's Fund; and Seth Berkley CEO of GAVI.[41]

The **BMGF** launched the **Coalition for Epidemic Preparedness Innovations (CEPI)** in 2017, to develop vaccines for emerging infectious diseases. The Scientific Advisory Committee includes Penny Heaton, from the Bill and Melinda Gates Medical Research Institute, and five non-voting members from 4 drug companies – Tekeda, Sanofi-Pasteur, Johnson and Johnson, and Pfizer.[42]

CEPI also has a **"Joint Coordination Group"** that is involved in research and development, regulation, stockpiling, and delivery of vaccines, as well as "planning for rapid response to a priority pathogen or unknown pathogen."[43] Members of this group include WHO, GAVI, EMA (European Medicines Agency), FDA, MSF, UNICEF, AFRC, AVAREF, NISBC, and Wellcome Trust.[44]

MSF (Medicins Sans Frontieres) is active in promoting vaccines. Its international medical coordinator, Myriam Henkens, says "The growth of the anti-vaccine movement in many developed countries seems absurd," and points out that these views are rare in the areas of the world in which her organization works.[45]

- AFRC (Air Force Reserve Command)
- AVAREF (The African Vaccine Regulatory Forum)
- NISBC (National Institute for Biological Standards and Control)

Now things get really complicated. **The Biomedical Advanced Research and Development Authority (BARDA)** was established in 2006 "to develop and procure medical countermeasures that address the public health and medical consequences of chemical, biological, radiological, and nuclear (CBRN) accidents, incidents and attacks, pandemic influenza, and emerging infectious diseases." BARDA supports the development of drugs and vaccines.[46]

CARB-X announced an investment of $500 million to accelerate the development of drugs and vaccines. Sources of funding include:

- BARDA
- Fauci's NIAID
- Wellcome Trust

And the **BMGF** is listed as an Alliance Partner.[47]

Both Inovio and Moderna have received funding from CEPI.[48] Inovio's website lists these partners:[49]

- BMGF
- Fauci's NIAID
- DARPA (see below)

The **BMGF** funds CEPI, which funds Inovio, which is also funded by **BMGF**.

In April 2020, Moderna was given $483 million in taxpayer funds by BARDA to develop a vaccine for COVID-19.[50] But Moderna was the beneficiary of even more largesse from the government. Fauci's NIAID Vaccine Research Center partnered with Moderna to conduct a clinical trial of its vaccine.[51] Additional funding was provided by **CEPI**,[52] which receives funding from the **BMGF**. And the **BMGF** provided a grant to Moderna for $20 million.[53] Moderna's website lists **BMGF** as a partner, along with BARDA.[54]

Moncef Slaoui, a former Moderna executive, was appointed to co-chair the White House coronavirus vaccine project called Operation Warp Speed. Slaoui said, shortly after his appointment in a Rose Garden event, "Mr. President, I have very recently seen early data from a clinical trial with a coronavirus vaccine. And this data made me feel even more confident that we will be able to deliver a few hundred million doses of vaccine by the end of 2020." Three days later, Moderna announced that early results from human trials for a coronavirus vaccine were promising.[55]

Yet another agency is the **Defense Advanced Research Projects Agency (DARPA),** a research branch of the Pentagon, which makes investments in technologies that promote national security.[56] DARPA researchers have been working on vaccine technology since the early 2010s. The agency awarded $25 million to Moderna for the development of mRNA vaccines which were the first to enter clinical trials. DARPA was also involved in developing COVID-19 tests.[57]

The Global Preparedness Monitoring Board (GPMB) was formed to ensure preparedness for global health crises. Its board members include:

- Dr. Victor Dzau, President of the National Academy of Medicine
- Dr. Chris Elias, President Global Development Program, **BMGF**
- Sir Jeremy Farrar, Director Wellcome Trust
- D. Anthony Fauci, Director National Institute of Allergy and Infectious Diseases
- George F Gao, Director-General Chinese Center for Disease Control and Prevention

GPMB lists as partners and founders The World Health Organization and the World Bank.[58] Who funds GPMB? Of course, the donor list includes the **BMGF**.[59]

The National Academy of Sciences was chartered in 1863 by Congress and President Lincoln to advise the government on science and technology. The Institute of Medicine (IOM) was founded in 1970 under the charter of the National Academy of Sciences to address medicine and healthcare and was an independent, non-profit, and non-governmental organization.[60] In 2015 the IOM became the National Academy of Medicine (NAM), and is now a private, non-profit institution that works outside government.[61]

Victor Dzau is President of NAM and the **BMGF** provides funding.[62]

The Centers for Disease Control's Advisory Committee on Immunization Practices (ACIP) has been issuing vaccine recommendations since 1995. ACIP recommendations are published by CDC and have a major impact on vaccine policy in U.S. and in other countries. ACIP is the "keys to the mandate kingdom" – used to get state legislators to pass laws mandating vaccines as a condition for admittance to school. The CDC is funded, in part, by the CDC Foundation. The **BMGF** is a donor to the Foundation.[63]

Research! America alliance and its member organizations endeavors to develop funding for medical and health research, inform the public, and motivate the public to support medical and health research.[64] Member organizations include almost two dozen pharmaceutical companies.[65] Victor Dzau is a former board member[66] and when "vaccinations" is entered into the website's search engine, 466 articles, notifications, and projects are listed.[67] Entering "Gates" in the search engine yielded 107 items, including announcements of funding from the **BMGF**, and awards given to Bill and Melinda Gates. Many of the items are vaccine-related.[68]

Cochrane, until recently, was an independent collaboration of researchers, health professionals, and patients who gathered and summarized the best evidence from published research in order to guide informed decision-making about healthcare matters. The organization prided itself on not accepting commercial funding and thus allow its reviewers to remain free of influence by commercial and financial interests.[69]

While the Cochrane website still shows that the organization is free from outside interest, this is no longer true. Cochrane accepted a donation of $1.5 million from the **BMGF** in 2016.[70] While this does not sound like a lot of money, it was enough money to start changing the organization's views on vaccines.

In fact, a 2018 article in the *British Medical Journal* called for a retraction of a Cochrane Review on the HPV Vaccine due to the influence of the foundation. A known advocate of the HPV vaccine, Lauri Markowitz, was involved in the Cochrane review. Markowitz had published articles promoting the vaccine and served on the US Advisory Committee on Immunization Practices (ACIP) Human Papillomavirus Working Group. She was the 'corresponding preparer' of the ACIP's documents for implementing HPV vaccination.[71]

In September 2018, Cochrane expelled a well-respected and long-time member, and director of the Nordic Cochrane Center in Copenhagen Denmark, Dr. Peter Gøtzsche, who had become outspoken about some vaccines, particularly the HPV vaccine. Gøtzsche, Cochrane member Tom Jefferson, and another colleague from the Collaboration had written a rebuttal to a paper from the Collaboration regarding safety of the vaccine, but the Collaboration dismissed their concerns and decided to oust Gøtzsche.[72]

Bill Gates managed to buy one of the few research organizations still willing to criticize vaccines, and not much money was required.

World Health Organization

Until the Trump Administration halted funding to the WHO, the United States was the largest single donor to the WHO, contributing over $400 million or 15% of the WHO's budget in 2019. At the time this book was being written, the **BMGF** was the largest donor, at over $200 million per year, or 9.8% of its funding.[73] The Gates Foundation has given over $2.4 billion to the WHO since 2000, and an article in Politico in 2017 titled, "Meet the world's most powerful doctor: Bill Gates" started with this observation: Some billionaires are satisfied with buying themselves an island. Bill Gates got a United Nations health agency in Geneva."[74]

Tedros Adhanom Ghebreyesus, Director General of the WHO

As previously mentioned, Ghebreyesus, the first African to head the World Health Organization, became Director-General in 2017. He is also the first person to head this organization who is not a physician. Prior to being elected to lead WHO, he was Ethiopia's Minister of Health from 2005-2012 and Minister of Foreign Affairs from 2012-2016. A commentary authored by Tedros on the WHO site includes this statement "All roads lead to universal health coverage for Dr. Tedros, and he has demonstrated what it takes to expand access to healthcare with limited resources."[75]

From 2009-2011, Ghebreyesus was also Director of the Global Fund, formed to fight AIDS, tuberculosis, and malaria, which was started by the **BMGF**. The Foundation committed $650 million to the Global Fund when it started, and another $750 million ten years later.[76] Ghebreyesus served on the Board of the GAVI Alliance for Immunization and has worked closely with the Clinton Foundation. While he was health minister of Ethiopia, The **BMGF** funded programs in his country, leading some people to think that Gates favored his election. Concern over Gates' influence on WHO led 30 advocacy groups to sign a letter to the WHO Executive Board stating their opposition to making the Gates Foundation an official partner of the agency.[77]

Two months before his election, Ghebreyesus was invited to deliver a keynote speech at Peking University, for an event at which Bill Gates also delivered a speech on "Looking to the Future: Innovation, Philanthropy, and Global Leadership." During his talk, Gates said, "With its rich pool of talented scientists and its capacity to develop new drugs and vaccines, China was a clear choice for us to locate a new Global Health Drug Discovery Institute. This institute—a collaboration between our foundation, the Beijing Municipal Government, and Tsinghua University—will help speed the discovery and development of new lifesaving medicines."[78] The Global Health Discovery Institute is involved in many projects, including the development of vaccines.[79]

Not surprisingly, in 2020 Tedros praised the Chinese government's response to the COVID-19 outbreak,[80] and thanked the Chinese government for its transparency.[81] He also spoke out against the U.S. and other countries that closed their borders after it was clear that China was not going to contain its spread.[82]

An Interesting Connection to the WHO

Margaret Chan, the former Director General of the WHO and who directed the declaration of the fake H1N1 pandemic, appointed Peng Liyuan as the WHO Goodwill Ambassador for Tuberculosis and HIV/AIDS. Ms. Peng is described on the WHO's website as a "famous Chinese soprano and actress," "head of the Chinese Song and Dance Ensemble in the General Political Department of the People's Liberation Army and ranked first class in the civil service with the military rank of major general."

The WHO site also states that "Ms. Peng is a strong advocate of health and the control of tuberculosis and HIV. In China she became the Minister of Health Ambassador for HIV/AIDS Prevention in January 2006 and the National Ambassador for TB Control and Prevention in March 2007."[83]

Tedros renewed her two-year term in 2019.

What the site does not say, is that Peng is the wife of China's President Xi Jinping. The site also does not list some of her major accomplishments as a member of the Chinese military. She joined the People's Liberation Army in 1980. She became famous as a star of China's state-run television network, singing songs that praised the Communist Party and celebrated her country's rise as a world power. In uniform, she sang for Chinese soldiers as they slaughtered pro-democracy protestors in Tiananmen Square.

Peng often accompanies her husband to key gatherings, including UN meetings. It was at one of these meetings where she met and became friendly with Bill Gates.

President Trump withdrew funding from the WHO, calling the organization a 'pipe organ' for Beijing's interest.[84]

More Collaboration

On May 14, 2018, Victor Dzau appeared at McGill University to inaugurate the McGill School of Population and Global Health. Tim Evans (formerly with WHO, Rockefeller Foundation, and GAVI) was installed as the Inaugural Director and Associate Dean of the school.[85] Paul Farmer delivered the inaugural lecture.[86] Farmer is the co-founder of **Partners in Health**, which is contracting with state and local governments to perform contact tracing as a means of stopping the spread of COVID-19.

Who funds **Partners in Health**? Of course, the Bill and Melinda Gates Foundation.[87] [88] [89]

Controlling the Message

To "sell" vaccines to the public, it is especially important to prevent free speech about vaccines, since a growing number of people are questioning vaccines and current recommendations. Censorship is accomplished in several ways, one of which is the International Fact-Checking Network. The Network states on its website that one of its activities is to "...cover trends, formats and news regarding fact-checking, "fake news" and misinformation."[90] Launched in 2015, the organization lists the **BMGF** as one of the organizations from which it has received funding.[91] Entering "vaccines" in the search engine leads to a page on which numerous critics and articles concerning some of the negative effects of vaccines are labeled "false."[92] All statements on the day we searched the site were labeled false, which either means that only those labeled "false" are posted, or all counter-arguments to vaccines are labeled false.

Are you confused? Most people are. But the important take-home point is that the **BMGF** has gained considerable influence over almost every important health-related agency worldwide by providing significant funding both directly and indirectly. The boards of these agencies are mostly populated with friends of Gates. The **BMGF** has gained control of world health and Bill Gates really has become the most powerful doctor in the world. And Gates knows it.

Many people have asked about Gate's intentions. We really do not know, and probably never will. Let us assume his intentions are good. They might be. The decision to use one's wealth for the betterment of society is a noble one, so perhaps Gates sees his massive investment in vaccines and technology as a good thing. But this does not mean that it is, and certainly there should be considerable public disclosure and debate about all of this. On the contrary, anyone who speaks out about it is silenced.

Was COVID-19 Planned?

You already read about the Rockefeller Foundation's scenario planning exercise. But there is more.

During a forum on pandemic preparedness at Georgetown University Dr. Anthony Fauci predicted that the Trump Administration would have to address a surprise disease outbreak, and that "risks have never been higher."[93] This was an amazing prediction based on the events of 2020.

The Global Preparedness Monitoring Board Annual Report 2019 warned that the world was at risk. "For too long, we have allowed a cycle of panic and neglect when it comes to pandemics: we ramp up efforts when there is a serious threat, then quickly forget about them when the threat subsides. It is well past time to act."[94]

The report includes "Seven Urgent Actions to Prepare the World for Health Emergencies"[95] The seventh one: The United Nations must strengthen coordination mechanisms. A list of "Progress indicator(s) by September 2020" included:

The United Nations (including WHO) conducts at least two system-wide training and simulation exercises, including one for covering the deliberate release of a lethal respiratory pathogen.[96]

The Johns Hopkins Center for Health Security works to protect people from epidemics and disasters. It receives funding from:

- WHO
- **BMGF**
- Rockefeller Foundation

- CDC
- US Department of State
- FDA

Event 201 was hosted by The Johns Hopkins Center for Health Security in partnership with the World Economic Forum and the **BMGF**. Described as a "high-level pandemic exercise," it took place on October 18, 2019 in New York and was a simulation of a coronavirus pandemic.[97]

Team leaders included:

- Eric Toner, Crystal Watson, Tara Kirk Sell from the Johns Hopkins Center for Health Security
- Ryan Morhard, World Economic Forum
- Jeffrey French, **BMGF**

The exercise team members for this event included:

- Dr. Christopher Elias, President of Global Development **BMGF**
- Dr. George Fu Gao, Director General, Chinese Center for Disease Control and Prevention
- Timothy Grant Evans, former Assistant Director at the World Health Organization, former Director of the Health Equity Theme at the Rockefeller Foundation, co-founder of GAVI
- Stephen C. Redd, Director of the CDC's Center for Preparedness and Response
- Jane Halton, Chairman of CEPI

The story line for Event 201 is that in late summer 2019, a virus infecting pigs in Brazil jumps to humans and by October starts to spread quickly enough that the world starts paying attention. The virus is a coronavirus called CAPS, which has never been seen before. It causes pneumonia and acute respiratory distress and in severe cases the lungs become filled with fluid and patients cannot breathe.

As the virus spreads, a Pandemic Response Board is formed with business leaders, public health experts and representatives from the Centers for Disease Control to develop and execute plans for stopping the spread of the virus. The fifteen people who were recruited to participate in this simulation are the members of the board and the Johns Hopkins' Center for Health Security conducted the exercise.

Like the news seen daily during the fake COVID-19 pandemic, increasing cases and deaths were shown on a screen, and ultimately 65 million people die. The world's GDP plunges by 11% and governments are paralyzed.

The exercise resulted in recommendations to address a future pandemic, including educating businesses about how devastating pandemics can be to the economy and the importance of contingency plans, building up stockpiles of medical supplies, and ensuring that adequate transportation is in place to deliver goods where needed.

There was also discussion about the importance of fighting fake news, conspiracy theories and propaganda campaigns deemed false by the committee. In the simulation, it was considered necessary to counter "fake news" but in the CAPs outbreak, the government overreaches and suppresses all citizens who speak out and all political opposition. Some governments tighten the reigns even more and institute martial law.

Jane Halton, Chairman of CEPI and an Event 201 player, expressed concern that the development of vaccines for emerging diseases is both time-consuming and expensive, and made the patently false statement that drug companies have little incentive to make such vaccines. Her statement was not challenged by anyone in the group.

The group issued final recommendations that included governments providing more resources and support for the development and manufacturing of vaccines, tests, and treatments; and the development of measures to combat misinformation and disinformation in response to pandemics. The group was quite specific in recommending partnerships with traditional and social media companies to counter misinformation, flooding the media with consistent information, and recruiting trusted and influential people to "readily and reliably augment public messaging, manage rumors and misinformation, and amplify credible information." The group also suggested that national public health officials should partner with the WHO to create and release consistent health messages; and that media companies should be asked to "...commit to ensuring that authoritative messages are prioritized and that false messages are suppressed through the use of technology."[98]

This exercise was a scenario exercise, like the one conducted by the Rockefeller Foundation in 2010. Is it just a coincidence that the COVID-19 debacle is eerily similar to both?

Gates denies that a simulation took place.

On April 12, Gates told BBC Breakfast "Now, here we are, we didn't simulate this, we didn't practice," he said. "So both in health policies and economic policies, we find ourselves in un-charted territory."[99]

Collectively, this information is provocative, and strongly suggests that the pandemic was not an accident. We will probably never know for sure. But as you will later read, these powerful people and organizations quickly mobilized to capitalize on the situation in order to promote agendas that they had all discussed publicly for many years – mandatory vaccinations and consider-ably more control over people's daily lives.

ENDNOTES

1. Birn AE, Fee E. The Rockefeller Foundation and the international health agenda. *Lancet.* 2013;381(9878):1618-1619. doi: https://doi.org/10.1016/S0140-6736(13)61013-2.

2. Ibid.

3. Galton F. EUGENICS: ITS DEFINITION, SCOPE AND AIMS. *Am J Soc.* 1904;10(1).

4. Messall R. The long road of eugenics: from Rockefeller to Roe v. Wade. *Hum Life Rev.* 2004;30(4):33-72.

5. Nicholas R. Scott. John D. Rockefeller Jr. & Eugenics: A Means of Social Manipulation. Page 19. https://www.yumpu.com/en/document/read/42433464/1-john-d-rockefeller-jr-eugenics-a-means-churchmilitant-tv. Accessed September 2, 2020.

6. Edwin Black. Eugenics and the Nazis – the California connection. SFGATE. https://www.sfgate.com/opinion/article/Eugenics-and-the-Nazis-the-California-2549771.php. November 9, 2003. Accessed May 17, 2020.

7. Stefan Kuhl. *The Nazi Connection: Eugenics, American Racism, and German National Socialism.* New York, NY: Oxford University Press;1994.

8. Ibid.

9. Lia Weintrab. The Link Between the Rockefeller Foundation and Racial Hygiene in Nazi Germany. Accessed September 2, 2020.

10. The Rockefeller Foundation. A Digital History. https://rockfound.rockarch.org/childhood-immunization. Accessed June 19, 2020.

11. Children's Vaccine Initiative. Grants RF 93059 and RF 94051, RAC, Unrpocessed, Box R3729, File 36019. The Rockefeller Foundation. https://rockfound.rockarch.org/digital-library-listing/-/

asset_publisher/yYxpQfeI4W8N/content/summary-of-grant-93059-to-children-s-vaccine-initiative. Accessed June 19, 2020.

12. Technology's Power to Transform the Lives of the Poor Revealed in New Study by the Rockefeller Foundation and Monitor's Global Business Network." *Business Wire.* June 21, 2010. https://www.businesswire.com/news/home/20100621005232/en/Technology%E2%80%99s-Power-Transform-Lives-Poor-Revealed-New. Accessed May 17, 2020.

13. Ibid.

14. The Rockefeller Foundation and Global Business Network. Scenarios for the Future of Technology and International Development. https://www.nommeraadio.ee/meedia/pdf/RRS/Rockefeller%20Foundation.pdf. Accessed September 2, 2020.

15. Books by Peter Schwartz. Goodreads. https://www.goodreads.com/author/list/31654.Peter_Schwartz. Accessed September 2, 2020.

16. Nossel GJ. "The Global Alliance for Vaccines and Immunization--a Millennial Challenge." *Nat Immun* 2000;1(1):5-8.

17. The Bill & Melinda Gates Foundation. Gavi. https://www.gavi.org/investing-gavi/funding/donor-profiles/bill-melinda-gates-foundation. Accessed May 23, 2020.

18. Gavi receives US$ 5 million from The Rockefeller Foundation. Gavi. https://www.gavi.org/news/media-room/gavi-receives-us-5-million-rockefeller-foundation#:~:text=Geneva%2C%201%20June%202020%20%E2%80%93%20Through,tools%20and%20innovative%20information%2Dsharing. Accessed September 2, 2020.

19. The U.S. and Gavi, the Vaccine Alliance. Global Health Policy. Kaiser Family Foundation. June 3, 2020. https://www.kff.org/global-health-policy/fact-sheet/the-u-s-and-gavi-the-vaccine-alliance/#:~:text=The%20U.S.%20is%20one%20of,FY%202019%2C%20and%20FY%202020. Accessed September 2, 2020.

20. Operating Model. Gavi. https://www.gavi.org/our-alliance/operating-model. Accessed May 23, 2020.

21. Industrialized country pharmaceutical industry. Gavi. https://www.gavi.org/operating-model/gavis-partnership-model/industralised-country-pharmaceutical-industry. Accessed May 23, 2020.

22. Governance. Gavi. https://www.gavi.org/our-alliance/governance. Accessed May 23, 2020.

23. Bill and Melinda Gates Pledge $10 Billion in Call for Decade of Vaccines. Bill and Melinda Gates Foundation. Press Room. https://www.gatesfoundation.org/Media-Center/Press-Releases/2010/01/Bill-and-Melinda-Gates-Pledge-$10-Billion-in-Call-for-Decade-of-Vaccines. Accessed May 23, 2020.

24. Release of the Global Vaccine Action Plan review and lessons learned report. World Health Organization. https://www.who.int/immunization/global_vaccine_action_plan/en/. Accessed June 20, 2020.

25. Bill & Melinda Gates Foundation. The Global Fund. https://www.theglobalfund.org/en/private-ngo-partners/resource-mobilization/bill-melinda-gates-foundation/#:~:text=The%20Gates%20Foundation%20has%20contributed,Replenishment%2C%20covering%202020%2D2022. Accessed June 20, 2020.

26. Pledges at Global Fund Sixth Replenishment Conference. Oct 9-10 2019 Lyon France. The Global Fund. https://www.theglobalfund.org/media/8882/replenishment_2019sixthreplenishmentconferencepledges_list_en.pdf?u=637278307830000000. Accessed June 20, 2020.

27. Gates Foundation Commits $760 Million to Global Fund. *PND by Candid*. October 11, 2019. https://philanthropynewsdigest.org/news/gates-foundation-commits-760-million-to-global-fund. Accessed June 20, 2020.

28. Members. The Global Fund. https://www.theglobalfund.org/en/board/members/. Accessed June 20, 2020.

29. Global Health Leaders Launch Decade of Vaccines Collaboration. Bill and Melinda Gates Foundation. Press Room. https://www.gatesfoundation.org/Media-Center/Press-Releases/2010/12/Global-Health-Leaders-Launch-Decade-of-Vaccines-Collaboration. Accessed June 20, 2020.

30. Strategic Advisory Group of Experts on Immunization (SAGE) Decade of Vaccines Working Group (March 2013- August 2020). World Health Organization. https://www.who.int/immunization/sage/sage_wg_decade_vaccines/en/ Accessed June 20, 2020.

31. National Institutes of Health. The NIH Director. New Phase of Cooperation Between NIH and the Bill and Melinda Gates Foundation. October 7, 2014. https://www.nih.gov/about-nih/who-we-are/nih-director/statements/new-phase-cooperation-between-nih-bill-melinda-gates-foundation. Accessed May 23, 2020.

32. Grants Map. Grand Challenges. https://grandchallenges.org/#/map. Accessed May 23, 2020.

33. Varmus H, Klausner R, Zerhouni E, Acharya T, Daar AS, Singer PA. Grand Challenges in Global Health. *Science* 2003;302(5644):389-399. doi: 10.1126/science.1091769.

34. Challenges. Global Grand Challenges. https://gcgh.grandchallenges.org/challenges?f%5b0%5d=field_initiative%3A37072&f%5b1%5d=field_grant_opp_open_dates%253Avalue%3A2003&f%5b2%5d=open_year%3A2003&items_per_page=25 Accessed May 23, 2020.

35. Sandi Doughton. After 10 years, few payoff from Gates' 'Grand Challenges.' *Seattle Times.* December 22, 2014. https://www.seattletimes.com/seattle-news/after-10-years-few-payoffs-from-gatesrsquo-lsquogrand-challengesrsquo/. Accessed September 2, 2020.

36. Awarded Grants. Global Grand Challenges. https://gcgh.grandchal-lenges.org/grants?f%5B0%5D=field_challenge%253Afield_initia-tive%3A37072&f%5B1%5D=funding_year%3A2020&f%5B2%5D=-funding_year%3A2019&items_per_page=100. Accessed May 23, 2020.

37. About us. Foundation for the National Institutes of Health. https://fnih.org/about. Accessed June 19, 2020.

38. FNIH 2017 Annual Report. Foundation for the National Institutes of Health. https://fnih.org/2017-annual-report/donors/. Accessed June 19, 2020.

39. Lev Facher. NIH partners with 16 drug companies in hopes of accelerating Covid-19 treatments and vaccines. *STAT.* April 17, 2020. https://www.statnews.com/2020/04/17/nih-partners-with-16-drug-companies-in-hopes-of-accelerating-covid-19-treatments-and-vaccines/. Accessed September 3, 2020.

40. Helen Branswell. With new grants, Gates Foundation takes an early step toward a universal flu vaccine. *STAT.* August 29, 2020. https://news.yahoo.com/grants-gates-foundation-takes-early-083548070.html. Accessed September 3, 2020.

41. Chan M, Elias C, Faauci A, Lake A, Berkley S. Reaching everyone, everywhere with life-saving vaccines. *Lancet.* 2017;389(1007):P777-779.

42. Who We Are. CEPI. https://cepi.net/about/whoweare/. Accessed May 23, 2020.

43. Ibid.

44. Ibid.

45. Overcoming barriers so vaccines can save lives. *Medicins Sans Frontieres.* https://www.msf.org/overcoming-barriers-so-vaccines-can-save-lives. Accessed June 19, 2020.

46. Introduction. Public Health Emergency. https://www.phe.gov/about/barda/stratplan/Pages/introduction.aspx#:~:text=The%20Pandemic%20and%20All%2DHazards,established%20BARDA%20in%20December%202006. Accessed June 19, 2020.

47. Funding Partners. CARB-X. https://carb-x.org/partners/funding-partners/. Accessed June 19, 2020.

48. Eric Sagonowsky. Inovio, Moderna score CEPI funding for vaccine work against deadly coronavirus. *Fierce Pharma.* January 23, 2020. https://www.fiercepharma.com/vaccines/inovio-moderna-score-cepi-funding-for-vaccine-work-against-deadly-coronavirus. Accessed September 3, 2020.

49. Partnerships. Inovio. https://www.inovio.com/about-inovio/partnerships/overview/. Accessed June 19, 2020.

50. David Mitchell, Sarah Kaminer Bourland. U.S. TAXPAYERS FUEL MODERNA'S COVID-19 VACCINE. *Patients for Affordable Drugs.* May 10, 2020. https://p4ad-main.friends.landslide.digital/2020/05/10/covid-blog-moderna/. Accessed September 3, 2020.

51. NIH Clinical Trial of Investigational Vaccine for COVID-19 Begins. March 16 2020 https://www.niaid.nih.gov/news-events/nih-clinical-trial-investigational-vaccine-covid-19-begins. Accessed September 3, 2020.

52. 52 David Mitchell, Sarah Kaminer Bourland. U.S. TAXPAYERS FUEL MODERNA'S COVID-19 VACCINE. *Patients for Affordable Drugs.* May 10, 2020. https://p4ad-main.friends.landslide.digital/2020/05/10/covid-blog-moderna/. Accessed September 3, 2020.

53. Foundations Advancing mRNA Science and Research. Moderna. https://www.modernatx.com/ecosystem/strategic-collaborators/foundations-advancing-mrna-science-and-research Accessed June 19, 2020.

54. Moderna's Company History. Moderna. https://www.modernatx. com/about-us/moderna-facts Accessed June 19, 2020.

55. Christina Wilkie. White House coronavirus vaccine advisor Moncef Slaoui to divest $12.4 million of Moderna holdings. *CNBC Politics.* May 18, 2020. http://www.cnbc.com/2020/05/18/coronavirus-vac-cine-adviser-moncef-slaoui-to-divest-12point4-million-of-moder-na-holdings.html. Accessed September 3, 2020.

56. About Darpa. Defense Advanced Research Projects Agency. https:// www.darpa.mil/about-us/about-darpa. Accessed June 19, 2020.

57. Andrew Eversden. How past investments positioned DARPA to take on coronavirus. *CYISRNET.* April 7, 2020. https://www.c4isrnet. com/industry/2020/04/07/how-past-investments-positioned-dar-pa-to-take-on-coronavirus/. Accessed September 3, 2020.

58. Board. Global Preparedness Monitoring Board. https://apps.who.int/ gpmb/board.html. Accessed May 23, 2020.

59. A World at Risk. Annual report on global preparedness for health emergencies. Global Preparedness Monitoring Board September 2019. https://apps.who.int/gpmb/assets/annual_report/GPMB_annu-alreport_2019.pdf. Accessed September 3, 2020.

60. Fallon H. The Institute of Medicine and Its Quality of Healthcare in America Reports. *Trans Am Clin Climatol Assoc* 2002;113:119-124.

61. History. National Academy of Sciences. http://www.nasonline.org/ about-nas/history/ Accessed May 23, 2020.

62. Giving to the National Academies. National Academy of Sciences https://www.nationalacademies.org/giving. Accessed June 19, 2020.

63. Our Partners: Foundations. CDC Foundation. https://www.cdcfoun-dation.org/partner-list/foundations. Accessed May 23, 2020.

64. About Us. Research!America. https://www.researchamerica.org/ about-us. Accessed May 23, 2020.

65. Member Organizations. Research!America. https://www.researcha-merica.org/about-us/member-organizations. Accessed September 3, 2020.

66. Research!America Board Member Victor Dzau Named Institute of Medicine President. Research!America. https://www.researchamer-ica.org/news-events/newsletter/researchamerica-board-member-vic-tor-dzau-named-institute-medicine-president. Accessed May 23, 2020.

67. Search: Vaccines. Research!America. https://www.researchamerica.org/search-site/vaccinations. Accessed May 23, 2020.

68. Search: Gates. https://www.researchamerica.org/search-site/Gates. Accessed May 23, 2020.

69. About Us. Cochrane. https://www.cochrane.org/about-us. Accessed September 3, 2020.

70. Cochrane announces support of new donor. Cochrane. https://www.cochrane.org/news/cochrane-announces-support-new-donor#:~:tex-t=COVID%2D19)%20resources-,Cochrane%20announces%20sup-port%20of%20new%20donor,on%20maternal%20and%20child%20health. Accessed September 3, 2020.

71. Hart EM. Rapid Response: Cochrane HPV vaccine review severely compromised by conflicts of interest. *BMJ* 2018:362:k3472.

72. Adam Marcus, Ivan Oransky. Turmoil erupts over expulsion of member from leading evidence-based medicine group. *STAT.* September 16, 2020. https://www.statnews.com/2018/09/16/expul-sion-cochrane-peter-gotzsche-medicine/. Accessed September 3, 2020.

73. Josephine Mouldes. How is the World Health Organization funded? *World Economic Forum.* April 15, 2020. https://www.weforum.org/agenda/2020/04/who-funds-world-health-organization-un-coronavi-rus-pandemic-covid-trump/. Accessed September 3, 2020.

74. Natalie Huet, Carmen Paun. Meet the world's most powerful doctor: Bill Gates. *Politico.* May 4, 2017. https://www.politico.eu/article/bill-gates-who-most-powerful-doctor/. Accessed September 3, 2020.

75. https://www.who.int/news-room/commentaries/detail/all-roads-lead-to-universal-health-coverage

76. Foundation Commits $750 Million to Global Fund. Bill and Melinda Gates Foundation. Bill & Melinda Gates Foundation. https://www.gatesfoundation.org/Media-Center/Press-Releases/2012/01/Foundation-Commits-$750-Million-to-Global-Fund. Accessed May 23, 2020.

77. Natalie Huet, Carmen Paun. Meet the world's most powerful doctor: Bill Gates. *Politico.* May 4, 2017. https://www.politico.eu/article/bill-gates-who-most-powerful-doctor/. Accessed September 3, 2020.

78. Bill Gates. My Advice for China's Students. GatesNotes. March 24, 2017. https://www.gatesnotes.com/Development/Peking-University-Speech Accessed May 23, 2020.

79. Case Highlights. Bill & Melinda Gates Foundation. https://www.gatesfoundation.org/Where-We-Work/China-Office/Case-Highlights. Accessed May 23, 2020.

80. President Xi Jinping: We will definitely overcome this disease. CGTN. YouTube. January 28, 2020. https://www.youtube.com/watch?v=6DiiY8a6xMg. Accessed September 3, 2020.

81. Barnini Chakraborty. China's relationship with WHO chief in wake of coronavirus outbreak under the microscope. *Fox News.* March 20, 2020. https://www.foxnews.com/world/coronavirus-china-who-chief-relationship-trouble. Accessed September 3, 2020.

82. Lisa Schlein. WHO Chief Urges Countries Not to Close Borders to Foreigners From China. *VOA.* Feb 3, 2020. https://www.voanews.com/science-health/coronavirus-outbreak/

who-chief-urges-countries-not-close-borders-foreigners-china.
Accessed September 3, 2020.

83. Goodwill ambassadors. World Health Organization. https://www.
who.int/about/who-we-are/structure/goodwill-ambassadors.
Accessed July 4, 2020.

84. Mail on Sunday Reporter. World Health Organisation fails to
mention its 'goodwill ambassador' Peng Liyuan is the wife of
China's President... saying only on its website that she's a singing star, amid concerns over WHO's handling of the coronavirus
pandemic. *Daily Mail.* May 23, 2020. https://www.dailymail.
co.uk/news/article-8351105/WHO-hails-goodwill-ambassador-
Peng-Liyuan-fails-mention-shes-wife-Chinas-President.html.
Accessed September 3, 2020.

85. Tim Evans. McGill International TB Center. https://www.mcgill.ca/
tb/investigators/tim-evans-0. Accessed May 23, 2020.

86. Dr. Paul Farmer visits McGill Uniersity as inaugural speaker to
kick-off new lecture series. *Med e-News. May 18, 2018. https://pub-
lications.mcgill.ca/medenews/2018/05/18/dr-paul-farmer-visits-mc-
gill-university-as-inaugural-speaker-to-kick-off-new-lecture-series/*
Accessed May 23, 2020.

87. Grant. Partners In Health A Nonprofit Corporation. Bill & Melinda
Gates Foundation. https://www.gatesfoundation.org/How-We-
Work/Quick-Links/Grants-Database/Grants/2014/10/OPP1120523.
Accessed September 3, 2020.

88. Grant. Partners In Health A Nonprofit Corporation. Bill & Melinda
Gates Foundation. https://www.gatesfoundation.org/How-We-
Work/Quick-Links/Grants-Database/Grants/2019/03/OPP1211539.
Accessed September 3, 2020.

89. Grant. Partners In Health A Nonprofit Corporation. Bill & Melinda
Gates Foundation. https://www.gatesfoundation.org/How-We-Work/

Quick-Links/Grants-Database/Grants/2016/06/OPP1131201.
Accessed September 3, 2020.

90. International Fact-Checking Network Transparency Statement.
Poynter. https://www.poynter.org/international-fact-checking-net-
work-transparency-statement/. Accessed September 3, 2020.

91. Ibid.

92. Search Results For: Vaccines. Poynter. https://www.poynter.org/?s=-
vaccines Accessed May 23, 2020.

93. Gerard Gallagher. Fauci: 'No doubt' Trump will face surprise infectious disease outbreak. *Healio Special Report: Health Care and
Politics.* January 11, 2017. https://www.healio.com/infectious-dis-
ease/emerging-diseases/news/online/%7B85a3f9c0-ed0a-4be8-9ca2-
8854b2be7d13%7D/fauci-no-doubt-trump-will-face-surprise-infec-
tious-disease-outbreak. Accessed May 23, 2020.

94. A World at Risk. Annual report on global preparedness for health
emergencies. Global Preparedness Monitoring Board September
2019. PDF File; 6. Phttps://apps.who.int/gpmb/assets/annual_report/
GPMB_annualreport_2019.pdf. Accessed September 3, 2020.

95. Ibid, 7.

96. Ibid, 39.

97. Event 201. Center for Health Security. https://www.centerfor-
healthsecurity.org/event201/about. Accessed May 23, 2020.

98. Event 201. A Global Pandemic Exercise. https://www.centerfor-
healthsecurity.org/event201/about. Accessed June 19, 2020.

99. Bill Gates: Few countries will get A-grade for coronavirus response. *BBC News.* April 12, 2020. https://www.bbc.com/news/av/
world-52233966/bill-gates-few-countries-will-get-a-grade-for-coronavi-
rus-response?fbclid=IwAR2NYrxTtvLavSU82tzSMGibcXXFQtowg-
DAEM-bOhBpbEamA_bV6I2RM9Gs Accessed May 23, 2020.

A QUESTIONABLE COLLABORATION

GAIN-OF-FUNCTION research involves manipulating viruses in a laboratory setting in order to investigate their potential to infect humans. This type of research is controversial due to the risk of accidental release of a mutated virus. While hundreds of researchers have spoken out against it, Dr. Fauci has historically defended this type of research. In an editorial in the *Washington Post* on December 30 2011, Fauci wrote: "[D]etermining the molecular Achilles' heel of these viruses can allow scientists to identify novel antiviral drug targets that could be used to prevent infection in those at risk or to better treat those who become infected. Decades of experience tells us that disseminating information gained through biomedical research to legitimate scientists and health officials provides a critical foundation for generating appropriate countermeasures and, ultimately, protecting the public health."[1]

Despite Fauci's enthusiasm for it, the National Institutes of Health issued a moratorium on funding for gain-of-function research in 2014. Researchers involved in this type of work were urged to discontinue their activities until risks and benefits could be more clearly defined.[2] A recent *Newsweek* article reports that reviews were conducted, although these were behind closed doors and away from public scrutiny. The moratorium was lifted in December 2017.

A new gain-of-function research project involving bat coronaviruses began in 2015, two years before the moratorium ended. Fauci's National Institute of Allergy and Infectious Diseases (NIAID) and the Chinese government authorized funding for both American researchers and the Wuhan Institute of Virology for the purpose of transforming a bat coronavirus into one that could infect and be transmitted by humans. They were successful, and the researchers reported their work in a prestigious European journal.[3] In the article, the researchers expressed some concern about whether their research was in violation of U.S. rules.

In 2019, the NIAID renewed the grant and committed an additional $3.7 million dollars for five more years of research, bringing the total invested in this research to $7.4 million. EcoHealth Alliance was the recipient of the grant. This organization describes itself as a "...global environmental health non-profit organization dedicated to protecting wildlife and public health from the emergence of disease." EcoHealth has some interesting partners which include:[4]

- Drug companies, including Johnson and Johnson, which has received hundreds of millions of dollars from the U.S. government for the development of a COVID-19 vaccine[5]

- Johns Hopkins School of Public Health, which staged Event 201 in October 2019, a simulation of a coronavirus pandemic that would kill 65 million people[6]
- The Centers for Disease Control and Prevention
- The National Institutes of Health
- The New York City Department of Health

The proposal for the more recent funding stated that, "We will use S protein sequence data, infectious clone technology, in vitro and in vivo infection experiments and analysis of receptor binding to test the hypothesis that % divergence thresholds in S protein sequences predict **spillover potential**."(emphasis ours)[7] "Spillover potential" means the ability of a virus to jump from animals to humans, and attach to receptors in human cells. The virus to be used in this research was a bat coronavirus.

The mainstream media has largely ignored any information that is critical of Fauci or the current response of government and health officials to COVID-19. An exception to this was a *Newsweek* article covering the story in late April.[8] *Newsweek* reported that Fauci did not respond to requests for comment, but that the NIH issued this statement: "Most emerging human viruses come from wildlife, and these represent a significant threat to public health and biosecurity in the US and globally, as demonstrated by the SARS epidemic of 2002-03, and the current COVID-19 pandemic.... scientific research indicates that there is no evidence that suggests the virus was created in a laboratory."[9]

The Trump administration ended funding for this research on April 24.[10]

At the time this book was being written, it was not known if Fauci had any direct involvement in arranging or overseeing these research projects. We do know that he was head of the NIAID during the entire time period in which the research was conducted. Agency heads are – or should be – held responsible for the actions of their employees and organizations. Generally, ignorance is not acceptable as a means for avoiding responsibility in many court proceedings. Perhaps other members of the mainstream media will become curious and decide to look into these matters further.

ENDNOTES

1. Anthony S. Fauci, Gary J. Nabel and Francis S. Collins. A flu virus risk worth taking. *Washington Post* December 30 2011 https://www.washingtonpost.com/opinions/a-flu-virus-risk-worth-taking/2011/12/30/gIQAM9sNRP_story.html accessed 9.1.2020

2. Akst J. "Moratorium on Gain-of-Function Research." *The Scientist* October 21 2014

3. Menachery VD, Yount Jr BI, Debbink K et al. "A SARS-like cluster of circulating bat coronaviruses shows potential for human emergence." *Nature Med* 2015 Nov:1508--1513

4. https://www.ecohealthalliance.org/partners accessed 9.1.2020

5. Schleunes A. "US Selects Two COVID-19 Vaccine Candidates for Huge Investment." *The Scientist* April 1 2020

6. http://www.centerforhealthsecurity.org/event201/ accessed 9.1.2020

7. Peter Daszak. Understanding the Risk of Bat Coronavirus Emergence. National Institutes of Health https://grantome.com/grant/NIH/R01-AI110964-06 accessed 9.1.2020

8. Fred Guterl. Dr. Fauci Backed Controversial Wuhan Lab With U.S. Dollars for Risky Coronavirus Research. *Newsweek* April 28 2020 accessed 9.1.2020

9. IBID

10. Sarah Owermohle . Trump cuts U.S. research on bat-human virus transmission over China ties. *Politico* April 27 2020 accessed 9.1.2020

CHINA BEFORE COVID-19

PRIOR TO THE COVID-19 debacle, China's economy had been growing at stellar rates. The goal of the Chinese Communist Party (CCP) was to surpass the United States as the number one superpower in the world. To accomplish this, they had opened their economy in a more capitalist manner, while remaining socially communist. They knew this was necessary to effectively compete in the world. The CCP also fueled rapid growth by manipulating its currency and establishing a massive trade deficit which was harmful to the United States. For decades, U.S. presidents did not hold China accountable, looking the other way while American companies exported businesses and jobs to China. All of this benefited the CCP. In addition, the CCP executed a policy of "forced technology transfer," which meant that all foreign companies doing business in China were forced to provide technology information to China in exchange for access to the Chinese market.[1] Many companies in the United States were more than happy to do this due to the sales potential they gained as a tradeoff. There have been many instances in which the CCP obtained technology due to this policy and patented it as its own.

Enter President Donald Trump. No president has been tougher on China since Richard Nixon. Trump campaigned heavily on the issue of holding the CCP accountable, and he explicitly stated that he considered China a threat to the U.S. economy and its safety. Immediately after becoming president, he began instituting measures to level the playing field with China. In 2018, he levied a 25% tariff on steel imports from China to the U.S., and a 10% tariff on aluminum imports. In 2018, the trade deficit with China was $419.5 billion,[2] which meant that the Chinese were exporting that much more to the U.S. than the amount of goods and services the U.S. was exporting to China. The deficit dropped to $345.6 billion by 2019, because of tariffs and other factors.

The CCP did not like this trend, as no president in recent decades had stood up to the regime in this way.

Another noteworthy issue is that the CCP has an abysmal human rights record, as is historically been the case with all communist regimes.[3] Most recently, their treatment of the Uighurs, a Muslim minority population in China, has been the subject of significant criticism from the world community. The atrocities inflicted on this population include forced sterilizations and abortions; home raids looking for copies of the Quran; and inspections in which children are counted followed by the imposition of unaffordable fines for having too many. Uighurs who do not pay the fines are taken to "re-education camps," which are in effect modern-day concentration camps. Although the CCP has relaxed its rules regarding having one child for the general Chinese population, the one child rule is still being enforced for the Uighurs, a thinly disguised program of eugenics and population control. The CCP has also destroyed hundreds of mosques in the region.[4] The CCP's goal is to destroy their identity and heritage, as they are a religious people, and religion is not compatible with communism.

Perhaps the most horrifying reports concerning the CCP's treatment of the Uighurs were those alleging that the CCP was using them for organ harvesting, an allegation that has been made for decades.[5] There have even been claims that healthy Uighurs were used for this purpose. Hamid Sabi, a lawyer for the China Tribunal has stated that he has proof that this was going on.[6]

These reports, combined with the fact that there were an estimated one million Uighurs and other Muslims in detention centers, made it increasingly more difficult for the world to ignore the situation. In the U.S., legislation was passed condemning the CCP and the surveillance and detention of this minority population.[7] Other countries followed suit.

Of course, the CCP denied this and stated that the detention centers were actually "labor and education camps," but they refused to allow any inspectors to see the facilities. The CCP was getting a lot of heat from around the world, and rightfully so.

Yet another issue concerning the CCP was its policies concerning Hong Kong, a long-time British colony that was returned to Chinese governance in 1997. The transfer was made under the premise that China would adhere to a "one country, two government systems" approach to Hong Kong. In other words, while technically a part of China, Hong Kong would remain a free area. This was the case for a long time, however the CCP, in conjunction with some local politicians, started working to gradually convert the island to communist status.

Citizens who were accustomed to freedom, and very aware and frightened of what life was like under the regime on the mainland, rebelled. Massive protests broke out, most significantly in 2019.[8] This was unacceptable to the CCP and they cracked down, sent in security forces and police, and used violence to stop the protestors.

In late November of 2019, President Trump signed into law the "Hong Kong Bill", which essentially recognized the city as autonomous from China and the CCP and stated that trade relations with Hong Kong would continue only as long as the island remained independent. The CCP vowed they would retaliate with "strong countermeasures."[9]

The CCP needed a distraction from the allegations of human rights violations and the increasing tensions in Hong Kong, and perhaps the answer was found with COVID-19. It was a marvelous way to sow chaos not only against its enemy, the U.S., but in the rest of the world as well. While the United States economy crashed, the Chinese economy surged ahead. President Xi just might have been thinking "mission accomplished."

ENDNOTES

1. Lee Branstetter What is the Problem of Forced Technology Transfer in China? *ECONOFACT* August 3 2018 https://econofact.org/what-is-the-problem-of-forced-technology-transfer-in-china accessed 9.1.2020

2. Kimberly Amadeo. US Trade Deficit With China and Why It's So High. The Real Reason Jobs Are Going to China. *World Economy Asia* February 26 2020 https://www.thebalance.com/u-s-china-trade-deficit-causes-effects-and-solutions-3306277 accessed 9.1.2020

3. China Events of 2018 https://www.hrw.org/world-report/2019/country-chapters/china-and-tibet# accessed 9.1.2020

4. Associated Press. China cuts Uighur births with IUDs, abortion, sterilization. June 29 2020 https://apnews.com/269b3de1af34e-17c1941a514f78d764c accessed 9.1.2020

5. Anastasia Lin. The ugly truth about China's organ harvesting. *New York Post* June 23 2019 https://nypost.com/2019/06/23/the-ugly-truth-about-chinas-organ-harvesting/ accessed 9.1.2020

6. Will Martin. China is harvesting thousands of human organs from its Uighur Muslin minority, UN human- rights body hears. *Business Insider* Sept 25 2019 https://www.businessinsider.com/china-harvesting-organs-of-uighur-muslims-china-tribunal-tells-un-2019-9?op=1 accessed 9.1.2020

7. Associated Press Posted by Shivani Kumar. US Congress approves China sanctions over ethnic crackdown. *Hindustan Times* May 28 2010 https://www.hindustantimes.com/world-news/us-congress-approves-china-sanctions-over-ethnic-crackdown/story-bEpSpR2W-sIaNfLIxaCIXHK.html accessed 9.1.2020

8. Verna Yu. Hong Kong: mammoth rally marks six months of pro-democracy protests. *The Guardian* December 8 2019 https://www.theguardian.com/world/2019/dec/08/

hong-kong-democracy-protests-continue-into-seventh-month accessed 9.1.2020

9. Fred Imbert China threatens to take 'strong counter-measures'
   against USS after Hong Kong bill signings. *CNBC* November 29
   2019 https://www.cnbc.com/2019/11/29/china-threatens-strong-
   counter-measures-after-hong-kong-bill-signings.html accessed
   9.1.2020

THE COVID-19 TIMELINE

2005

THE INTERNATIONAL Health Regulations (IHR) was agreed to by 196 countries including World Health Organization Member States for the purpose of working together for global health security. The IHR provides an "overarching legal framework that defines countries' rights and obligations in handling health events and emergencies that have the potential to cross borders. The Regulations also outline the criteria to determine whether an event constitutes a "public health emergency of international concern."[1]

2010

The Rockefeller Foundation published a report in partnership with the Global Business Network titled "Scenarios for the Future of Technology and International Development."[2] The collaboration used "scenario planning" to look at possible responses to hypothetical situations, including a pandemic. A scenario titled

"LOCK STEP" describes a world of tighter top-down government control and more authoritarian leadership with innovation and growing citizen pushback after a pandemic is declared. The events described in this report are eerily similar to what starts taking place in 2020.

2015

Gain-of-function research initiated, funded by the U.S. government. American researchers and the Wuhan Institute of Virology were tasked with transforming a bat coronavirus into one that could infect and be transmitted by humans. They were successful, and the researchers reported their work in a prestigious European journal.[3]

January 11, 2017

During a forum on pandemic preparedness at Georgetown University, Dr. Anthony Fauci predicted that the Trump Administration would have to address a surprise disease outbreak, and that "risks have never been higher."[4] This was an amazing prediction based on the events of 2020.

January 17, 2017 World Economic Forum Davos Meeting

Bill Gates announced a new working group:

Coalition for Epidemic Preparedness Innovations (CEPI)

This was a collaboration that included the Bill and Melinda Gates Foundation, governments of Norway, India, Japan, Germany, several drug companies, the Defense Advance Research Projects Agency (DARPA) and the U.S. Army Medical Research Institute for Infectious Diseases at Fort Dietrich, MD.[5]

April 27, 2018

During a discussion hosted by the Massachusetts Medical Society and the *New England Journal of Medicine* Bill Gates warns that a disease coming soon could kill 30 million people within 6 months.[6]

August 2019

The Bill and Melinda Gates Foundation helped to negotiate a $100 billion contract for population contact tracing six months before the COVID-19 pandemic began. Two investigators who testified in front of a Congressional hearing concerning the Clinton Foundation and tax fraud in 2018, John Moynihan and Larry Doyle, broke the story.

Representatives from the Gates Foundation met with Congressman Bobbie Rush in Rwanda in August 2019 to discuss who would get the contract for a massive government funded contact tracing program. Nine months later, Bobbie Rush, a Democrat from Illinois, introduced H.R. 6666, the COVID-19 Testing, Reaching and Contacting Everyone (TRACE) Act. This bill allocates $100 billion for this program, to be administered by the CDC and executed by the Rockefeller Foundation. Rush traveled to Rwanda with his wife for this meeting, and his trip was paid for by the Gates Foundation.[7]

October 18, 2019

A pandemic exercise called Event 201 was conducted at Johns Hopkins University, sponsored by the Bill and Melinda Gates Foundation, and the Michael Bloomberg School of Public Health at Johns Hopkins University. This simulation predicted that a coronavirus would have the same kill rate as Spanish Flu of 1918 which caused 50 million deaths worldwide in 18 months. Dr. George F Gao, Director of the Chinese Center for Disease Control and Prevention, was involved in the simulation.[8]

October 18, 2019

The World Military Games began in Wuhan China with almost 10,000 military personnel from over 110 countries, including the U.S. [9]

November 2019

A documentary called *The Next Pandemic* featuring Bill Gates was released. In the film, Gates predicted a virus outbreak would begin in a wet market in China.[10]

December 10, 2019

One of the earliest known COVID-19 patients starts feeling ill and is admitted to the Wuhan Central Hospital on December 16, 2019.[11]

January 8, 2020

Netflix releases a docuseries called *Pandemic: How to Prevent an Outbreak.* "I think the series couldn't have come at a more crucial time with the recent COVID-19 outbreak," said producer Mando Stathi. The docuseries features Bill Gates, and Dr. Syra Madad, the Senior Director of New York City's Health System, who later says, "We are all in it together."[12]

January 2020

The CDC decided to develop its own COVID-19 test.

January 23, 2020

Moderna Inc. announced a collaboration with CEPI and The Vaccine Research Center of the National Institute of Allergy and Infectious Diseases (Fauci's agency) to develop an mRNA vaccine against COVID-19. At the same time, Moderna and Inovio announced that they were working with the National Institutes of Health to develop a vaccine. The NIAID announced that it would allow the companies to bypass animal tests and proceed directly to human tests.[13]

Gilead announced that it would begin researching the potential for remdesivir, a drug that proved to be useless and harmful for treating Ebola, for the treatment of COVID-19.[14]

In an article published in the *Journal of the American Medical Association,* Fauci cited Gilead's drug remdesivir as a promising treatment for COVID-19.[15]

According to the WHO there were 581 confirmed cases of COVID-19 worldwide on January 23.[16] There was one patient quarantined in Washington State. Mr. Fauci, the drug companies, and CEPI seem to have an amazing ability to predict the future need for treatments and vaccines.

January 31, 2020

President Trump orders a travel ban from China.[17]

February 2020

CDC-developed COVID-19 test was distributed to health centers throughout the U.S., and within a few days, the tests were found to be inaccurate. The FDA insisted that hospitals, academic centers, and private companies *should not develop their own tests.*

Late February 2020

FDA finally lifted the ban on test development. In the rush to get tests ready for market the FDA provided no standards for how COVID-19 was supposed to be detected. This meant all test makers could use any standard they wanted to.

The tests were approved by the FDA under emergency use authorization, which means that they were only required to perform well in test tubes. No real world demonstration of clinical viability was required, according to David Pride MD, associate director of microbiology at the University of California San Diego.[18]

March 2020

CDC stated that test results are unreliable:

> "Results are for the identification of 2019-nCoV RNA. The 2019-nCoV RNA is generally detectable in upper and lower respiratory specimens during infection. Positive results are indicative of active infection with 2019-nCoV but do not rule out bacterial infection or co-infection with

other viruses. The agent detected may not be the definite cause of disease. Laboratories within the United States and its territories are required to report all positive results to the appropriate public health authorities."

"Negative results do not preclude 2019-nCoV infection and should not be used as the sole basis for treatment or other patient management decisions. Negative results must be combined with clinical observations, patient history, and epidemiological information."[19]

March 2020

Fauci reported in an article in the *New England Medical Journal* that "…the case fatality rate may be considerably less than 1%. This suggested that the overall clinical consequences of Covid-19 may ultimately be more akin to those of a severe seasonal influenza (which has a case fatality rate of approximately 0.1%)…"[20]

March 3, 2020

Fauci reported it will take 12-18 months to develop a vaccine for COVID-19.[21]

March 11, 2020

Fauci reported that the COVID-19 mortality rate was "ten times worse" than seasonal flu.[22] He told a Congressional hearing that "The flu has a mortality rate of 0.1 percent. This has a mortality rate of 10 times that. That's the reason I want to emphasize we have to stay ahead of the game in preventing this."[23]

March 11, 2020

The WHO declared COVID-19 a pandemic.[24]

Of note, at one time this was how the WHO defined a pandemic:

> "An influenza pandemic occurs when a new influenza virus appears against which the human population has no immunity, resulting in several, **simultaneous epidemics worldwide with enormous numbers of deaths and illness.** With the increase in global transport and communications, as well as urbanization and overcrowded conditions, epidemics due the new influenza virus are likely to quickly take hold around the world."

Severity is an issue since every year the seasonal flu causes a "global spread of disease."[25]

In 2009, the WHO changed the definition in order to declare H1N1 a pandemic. According to WHO Director-General Margaret Chan, "ministers of health" should take advantage of the "devastating impact" swine flu will have on poorer nations to get out the message that "changes in the functioning of the global economy" are needed to "distribute wealth on the basis of" values "like community, solidarity, equity and social justice."

She further declared the pandemic should be used as a weapon against "international policies and systems that govern financial markets, economies, commerce, trade and foreign affairs."[26]

"Pandemic" is a loosely defined term, although it is consistently effective at generating fear for the world's populations.

March 13, 2020

President Trump declared a national emergency, stating that this would allow him to send $50 billion to the states to help fight the disease. Broad new authority was granted to the secretary of Health and Human Services, and massive testing would begin soon. According to Fauci, Trump's action was a "forward-leaning" approach to the crisis. He said, "We still have a long way to go. There will be many more cases. But what's going on here today is going to help it to end sooner than it would have."[27]

There were 338 new COVID-19 cases reported in the U.S. on March 14, 2020.[28]

Mid-March 2020

Lockdowns began. The economy crashed. Tens of millions of people lost their jobs. The government took charge of private industry to increase production of certain products. The government decided which businesses could remain open and which would be forced to close.

Martial law was not required because 330,000,000 people voluntarily gave up their civil liberties, which included the right to assemble, and the right to worship.

The government warned people that using cash was not advisable, and some stores stopped accepting cash. Of note, cash is not trackable, unlike credit card and debit card transactions.

March 16, 2020

A Phase I clinical trial for a COVID-19 vaccine began. The first patient receives a vaccine called mRNA-1273 which was

119

developed by NIAID (Fauci's NIH agency) and Moderna with financial support from CEPI.[29]

A model developed by Neil Ferguson of the Imperial College of London predicted that tens of millions of people would die due to COVID-19 infection. COVID-19 was compared to the Spanish flu, which killed approximately 50 million people in 1918. Ferguson's report stated that the only way to prevent massive deaths would be for the entire population of the planet to be locked down and for people to remain separated for 18 months until a vaccine was available. Total isolation would be needed because the isolation of vulnerable populations like the elderly would only reduce deaths by half.[30]

March 30, 2020

Fauci told CNN that there would be between 100,000 and 200,000 deaths from COVID-19 in the U.S.[31] Dr. Deborah Birx agreed, calling this "our real number."

March 31, 2020

Gates wrote in an op-ed in the Washington Post that the U.S. missed the opportunity to get ahead of COVID-19. Claiming to have "spoken with experts and leaders in Washington and across the country. It's become clear to me that we must take three steps."

Shutdown anywhere means shutdown everywhere – at least 20 weeks and maybe more.

More testing.

Investment in treatments and a vaccine, estimated to take 18 months. Billions of doses will be needed to vaccinate the world's population.[32]

April 7, 2020

Dr. Deborah Birx announced that death certificates for anyone who dies with COVID-19 should reflect death by COVID-19 even if COVID-19 is not the cause of death.[33]

April 9, 2020

Fauci told Americans that the death toll would be more like 60,000.[34] This still sounds serious, but data for seasonal flu indicates that COVID-19 may be less deadly than we have been told.

July 3, 2020

Cases total 2,789,678 and deaths total 129,305, according to the CDC.[35] These numbers are likely significantly inflated based on the inaccuracy of the tests and inaccurate data entered on death certificates. In most areas of the U.S. there is no sign that normalcy will return any time soon.

ENDNOTES

1. International Health Regulations. World Health Organization. https://www.who.int/health-topics/international-health-regula-tions#tab=tab_1 accessed 9.1.2020

2. Technology's Power to Transform the Lives of the Poor Revealed in New Study by the Rockefeller Foundation and Monitor's Global Business Network." *Businesswire* June 21 2010 https://www.businesswire.com/news/home/20100621005232/en/Technology%E2%80%99s-Power-Transform-Lives-Poor-Revealed-New accessed 5.17.2020

3. Menachery VD, Yount Jr BI, Debbink K et al. "A SARS-like cluster of circulating bat coronaviruses shows potential for human emergence." *Nature Med* 2015 Nov;21:1508--1513

4. Gerard Gallagher. Fauci: 'No doubt' Trump will face surprise infectious disease outbreak. *Healio Special Report: Health Care and Politics* Jan 11 2017 https://www.healio.com/infectious-disease/emerging-diseases/news/online/%7B85a3f9c0-ed0a-4be8-9ca2-8854b2be7d13%7D/fauci-no-doubt-trump-will-face-surprise-infectious-disease-outbreak accessed 5.23.2020

5. CEPI. A global coalition for a global problem. https://cepi.net/about/whoweare/ accessed 5.23.2020

6. Kevin Loria. Bill Gates thinks a coming disease could kill 30 million people within 6 months – and says we should prepare for it as we do for war. *Business Insider* April 27 2018 https://www.businessinsider.com/bill-gates-warns-the-next-pandemic-disease-is-coming-2018-4?r=UK accessed 9.1.2020

7. Natural News. Six months before the pandemic Bill Gates negotiated a $100 million dollar contact tracing deal with a democratic congressman. *Newsbreak* https://www.newsbreak.com/news/1597065528531/

six-months-before-the-pandemic-bill-gates-negotiated-a-100-million-contact-tracing-deal-with-a-democratic-congressman accessed 9.1.2020

8. Event 201. A Global Pandemic Exercise. https://www.centerfor-healthsecurity.org/event201/about accessed 6.19.2020

9. Organizing Committee of the 7th CISM Military World Games. 2019 Military World Games kicks off in Central China's Wuhan. *PR Newswire* October 17 2019 https://www.prnewswire.com/news-re-leases/2019-military-world-games-kicks-off-in-central-chinas-wu-han-300940464.html accessed 9.1.2020

10. Corazon Miller. Bill Gates predicted a coronavirus-like outbreak – down to it starting at a Chinese market – in 2019 Netflix docu-mentary show 'The Next Pandemic.' *Daily Mail* Jan 31 2020 https://www.dailymail.co.uk/news/article-7951293/Bill-Gates-Predicted-Coronavirus-Like-Outbreak-2019-Netflix-Documentary.html accessed 9.1.2020

11. Bethany Allen-Ebrahimian. Timeline: The early days of China's coronavirus outbreak and cover-up. *Axios* Mar 18 2020 https://www.axios.com/timeline-the-early-days-of-chinas-coronavirus-out-break-and-cover-up-ee65211a-afb6-4641-97b8-353718a5faab.html accessed 9.1.2020

12. Chrissa Loukas. "Review: Pandemic How to Prevent an Outbreak." *The Corsair* Mar 26 2020 https://www.thecorsaironline.com/cor-sair/2020/3/26/pandemic-how-to-prevent-an-outbreak accessed 9.1.2020

13. Andrew Dunn. A coalition backed by Bill Gates is funding biotechs that are scrambling to develop vaccines for the deadly Wuhan coro-navirus. *Business Insider* Jan 23 2020 https://amp.businessinsider.com/vaccines-for-wuhan-china-cornonavirus-moderna-inovio-ce-pi-2020-1 accessed 9.1.2020

14. Gilead Assesses Ebola Drug as Possible Coronavirus Treatment. *Bloomberg Law* Jan 23 2020 https://news.bloomberglaw.com/pharma-and-life-sciences/gilead-assesses-ebola-drug-as-possible-coronavirus-treatment accessed 9.1.2020

15. Paules CI, Marston HD, Fauci AS. "Coronavirus Infections – More than Just the Common Cold." *JAMA.* 2020;323(8):707-708

16. World Health Organization. Novel Coronavirus (2019-nCoV) SITUATION REPORT-3 23 January 2020 https://www.who.int/docs/default-source/coronaviruse/situation-reports/20200123-sitrep-3-2019-ncov.pdf?sfvrsn=d6d23643_8 accessed 9.1.2020

17. Proclamations. Proclamation on Suspension of Entry of Immigrants and Nonimmigrants of Persons who Pose a Risk of Transmitting 2019 Novel Coronavirus. January 31 2020 https://www.whitehouse.gov/presidential-actions/proclamation-suspension-entry-immigrants-nonimmigrants-persons-pose-risk-transmitting-2019-novel-coronavirus/ accessed 7.4.2020

18. David Pride. Hundreds of different coronavirus tests are being used – which is best? *The Conversation* April 4 2020 https://www.marketwatch.com/story/hundreds-of-different-coronavirus-tests-are-being-used-which-is-best-2020-04-02 accessed 9.1.2020

19. CDC 2019-Novel Coronavirus (2019-nCoV) Real-Time RT-PCR Diagnostic Panel Instructions for Use Centers for Disease Control https://www.fda.gov/media/134922/download accessed 9.1.2020

20. Fauci AS, Lane HC, Redfield RR. "Covid-19 – Navigating the Uncharted." *NEJM* 2020 Mar;382:1268-1269

21. Stephanie Soucheray. Fauci: Vaccine at least a year away, as COVID-19 death toll rises to 9 in Seattle." *CIDRAP* Mar 3 2020 https://www.cidrap.umn.edu/news-perspective/2020/03/fauci-vaccine-least-year-away-covid-19-death-toll-rises-9-seattle accessed 9.1.2020

22. Ronald Bailey. COVID-19 Mortality Rate 'Ten Times Worse' Than Seasonal Flu, Says Dr. Anthony Fauci. *Reason* Mar 11 2020 https://reason.com/2020/03/11/covid-19-mortality-rate-ten-times-worse-than-seasonal-flu-says-dr-anthony-fauci/ accessed 9.1.2020

23. Joseph Guzman Coronavirus 10 times more lethal than seasonal flu, top health official says. *The Hill* https://thehill.com/changing-america/well-being/prevention-cures/487086-coronavirus-10-times-more-lethal-than-seasonal accessed 9.1.2020

24. WHO Director-General's opening remarks at the media briefing on COVID-19- 11 March 2020 https://www.who.int/dg/speeches/detail/who-director-general-s-opening-remarks-at-the-media-briefing-on-covid-19---11-march-2020 accessed 9.1.2020

25. World Health Organization. Communicable Disease Surveillance and Response (CSR). Pandemic preparedness. http://web.archive.org/web/20050210210053/www.who.int/csr/disease/influenza/pandemic/en/ accessed 9.1.2020

26. Director-General Address to the Regional Committee for Europe (59th Session). World Health Organization 15 September 2009 https://www.who.int/dg/speeches/2009/euro_regional_committee_20090815/en/ accessed 9.1.2020

27. Libby Cathey. Government coronavirus response; Trump declares national emergency, says he likely will get tested. *ABC News* Mar 13 2020 https://abcnews.go.com/Politics/government-coronavirus-response-live-updates-trump-declares-national/story?id=69580277 accessed 9.1.2020

28. CDC COVID Data Tracker Centers for Disease Control https://www.cdc.gov/coronavirus/2019-ncov/cases-updates/cases-in-us.html accessed 7.4.2020

29. NIH clinical trial of investigational vaccine for COVID begins. NIH News Releases. National Institutes of Health https://www.nih.gov/news-events/news-releases/

nih-clinical-trial-investigational-vaccine-covid-19-begins accessed 7 4 2020

30. Ferguson NM, Laydon D, Nedjati-Gilani G et al. "Report 9: Impact of non-pharmaceutical interventions (NPIs) to reduce COVID-19 mortality and healthcare demand." *Imperial College COVID-19 Response Team* March 16 2020 https://www.imperial.ac.uk/mrc-global-infectious-disease-analysis/covid-19/report-9-impact-of-npis-on-covid-19/ accessed 9.1.2020

31. Devan Cole, Kevin Bohn, Dana Bash. US could see millions of coronavirus cases and 100,000 or more deaths, Fauci says. *CNN* March 30, 2020 https://www.cnn.com/2020/03/29/politics/coronavirus-deaths-cases-anthony-fauci-cnntv/index.html accessed 9.1.2020

32. Bill Gates. Bill Gates: Here's how to make up for lost time on covid-19. *Washington Post* March 31 2020 https://www.washingtonpost.com/opinions/bill-gates-heres-how-to-make-up-for-lost-time-on-covid-19/2020/03/31/ab5c3cf2-738c-11ea-85cb-8670579b863d_story.html accessed 9.1.2020

33. Louis Casiano. Birx says government is classifying all deaths of patients with coronavirus as COVID-19 regardless of cause. *Fox News* April 7 2020 https://www.foxnews.com/politics/birx-says-government-is-classifying-all-deaths-of-patients-with-coronavirus-as-covid-19-deaths-regardless-of-cause accessed 9.1.2020

34. Ian Schwartz. Dr. Fauci: Virus Death Toll May Be "More Like 60,000 Than 100,000 to 200,000. *Real Clear Politics* April 9 2020 https://www.realclearpolitics.com/video/2020/04/09/dr_fauci_virus_death_toll_may_be_more_like_60000_than_100000_to_200000.html accessed 9.1.2020

35. CDC COVID Data Tracker. Centers for Disease Control. https://www.cdc.gov/coronavirus/2019-ncov/cases-updates/us-cases-deaths.html accessed 7.3.2020

HOW TO INVENT A PANDEMIC

THERE ARE DIFFERING stories about how the COVID-19 debacle began. What is clear is that when the dates of various events are matched with data on cases and deaths, the responses make no sense at all.

The official narrative is that in December 2019, China reported the first case. On December 31, 2019, Chinese health authorities reported that dozens of people in Wuhan, China were being treated for "pneumonia of unknown etiology." By Jan 3, 2020, Chinese officials reported a total of 44 patients to the World Health Organization (WHO). Of those 44, 11 were reported to be severely ill and the rest in stable condition. The Chinese stated that a wet market in Wuhan was the source of the disease and had been closed on January 1 for a thorough cleaning and disinfection.[1]

On January 11, 2020 China reported the first death – a 61-year-old man who had reportedly visited the wet market in Wuhan.[2]

The first cases outside China were reported on January 20, 2020 in Japan and South Korea. One day later the first case was reported in Washington State, according to the Centers for Disease Control. The infected man had traveled to Wuhan but stated that he had not visited the animal market. Governor Jay Inslee said, "This is certainly not a moment for panic or high anxiety...The risk is low to residents in Washington."[3]

On January 23, 2020 China reported locking down Wuhan, a city of 11 million people. All transportation into and out of the city was stopped including airports, ferries, and trains. Subways and buses were stopped. People panicked and soon the shelves of grocery stores and pharmacies were empty. At the time of the lockdown there were 571 cases reported worldwide.[4]

On January 29, 2020 President Trump announced that he was forming a coronavirus task force comprised of experts in health, transportation and national security.[5] There were a total of 91 confirmed cases outside China including 5 in the U.S.[6]

On January 30, 2020, The World Health Organization declared a "public emergency of international concern." At the time, there were 9700 cases of the disease in China and 106 cases in other countries.[7] The first person-to-person case in the U.S. was reported to occur between a husband and wife in Illinois.[8]

On February 5, 2020, over 3600 passengers on the Diamond Princess Cruise ship were quarantined off the coast of Japan. Eventually over 700 cases would be confirmed. On that day the 12th case of the disease was confirmed in the U.S. Two charter flights from Wuhan with 350 passengers landed at Travis Air Force Base in California. All travelers were placed in a 14-day quarantine.[9]

On February 11, 2020 WHO announced the official name of this new virus – COVID-19.[10]

On February 29, 2020, the first death was reported in Washington state.[11]

On March 3, 2020, the CDC issued guidance allowing anyone to be tested for COVID-19.

On March 11, 2020 the World Heath Organization declared COVID-19 a pandemic. On that day the U.S. had 29 deaths. China reported 24 cases that week. WHO head Tedros reported that 90% of cases were reported in only four countries.[12]

On March 13, 2020, Trump declared a national emergency. This action transferred broad new authority to the Secretary of Health and Human Services. Fauci stated at a press conference with Trump, "…we are bringing a whole of government approach to confronting the coronavirus…" Nancy Pelosi announced plans to pass a bill that would include free tests for everyone, paid sick and emergency leave, and enhanced unemployment benefits.[13] On this day there were 1678 cases and 41 deaths in the U.S.[14]

By March 17, 2020, there were cases in all 50 states, and residents of Northern California were ordered to "shelter in place" for three weeks. Colorado closed bars and eliminated dining in restaurants, and McDonald's announced that dining rooms would close in all company-owned stores. Many other states and municipalities started ordering closures, lockdowns and other restrictions, along with the governments of many other countries. There were 4661 cases in the U.S. Globally there were 181,580 cases and 7130 deaths.[15]

On March 18, 2020, China reported no new infections. Wuhan was the only city in Hubei province still in lockdown.[16]

On March 20, 2020, New York City was declared an outbreak epicenter by Mayor Bill de Blasio.[17] There were 5600 cases in New York City at the time.[18]

Lockdowns and closures continued, and many states extended their orders for additional weeks and then months.

On July 11, 2020, there were 3,173,212 cases and 133,486 deaths in the U.S.[19] There were 12,498,458 cases and 229,949 deaths worldwide.[20]

Was the Response Justified?

There are good reasons to believe that the data reported here are not valid and likely highly inflated. As you read earlier, according to the Centers for Disease Control (CDC)'s website, about 9% of the world's population is affected by flu annually with up to one billion infections, 3-5 million severe cases, and 300,000-500,000 deaths per year.[21] [22] It is estimated that 20% of Americans are affected, with 25-50 million documented cases, 225,000 hospitalizations and tens of thousands of deaths annually.[23] [24] [25] [26] [27] Historically, the elderly account for 90% of influenza deaths.[28] These data are for "normal" years. Based on these data, the response to COVID-19 was clearly disproportionate.

How The Fictional Tale Began

The fiasco started with a model developed by Neil Ferguson of the Imperial College of London which predicted that tens of millions of people would die due to COVID-19 infection. COVID-19 was compared to the Spanish flu, which killed approximately 50 million people in 1918. As previously stated, Ferguson's report stated that the only way to prevent massive deaths would be for the entire population of the planet to be locked down and for people to remain separated for 18 months until a vaccine was available. Total isolation would be needed because the isolation of just vulnerable populations like the elderly would only reduce deaths by half.[29]

Apparently "renowned experts" like Mr. Fauci and Deborah Birx did not check Ferguson's background. He had demonstrated on numerous occasions that he was unable to accurately predict anything. In 2002, he predicted that 150,000 people would die from Mad Cow Disease, but only 2704 died. His estimation was 55 times higher than the real number. A few years later he predicted that 65,000 people would die of swine flu, and only 457 people died – his estimation was 142 times higher than the real number.[30] [31] And his prediction of deaths from bird flu was 200,000,000 and only 455 people died – a prediction 439,560 times higher than the real number.[32]

As of June 25, 2020, the total deaths worldwide from COVID-19 had reached 494,179 – not tens of millions – and even this number is questionable. This time Ferguson was off not by thousands or hundreds of thousands, but by millions. Also, the average age at death was 80, with almost all who died having multiple co-morbidities. The virus has had little effect on young, healthy people.

Enter Fauci: Liar in Chief

Mr. Fauci reported in an article in the *New England Medical Journal* published in March 2020 that "…the case fatality rate may be considerably less than 1%. This suggests that the overall clinical consequences of Covid-19 may ultimately be more akin to those of a severe seasonal influenza (which has a case fatality rate of approximately 0.1%)."[33]

Yet just days later, on March 11, 2020 Fauci said that the COVID-19 mortality rate was "ten times worse" than seasonal flu.[34] At a Congressional hearing on March 11, 2020, Fauci said that "The flu has a mortality rate of 0.1 percent. This has a mortality rate of 10 times that. That's the reason I want to emphasize we have to stay ahead of the game in preventing this."[35] Both of Fauci's statements could not be true – COVID-19 cannot be similar to normal seasonal flu AND have a death rate 10 times higher than seasonal flu.

WHO Director-General Tedros also fanned the fictional flames, stating on March 4, 2020, "Globally, about 3.4% of reported COVID-19 (the disease spread by the virus) cases have died. By comparison, seasonal flu generally kills far fewer than 1% of those infected."[36]

President Trump disagreed, stating that he had consulted with experts who said that many people who are exposed to flu are either asymptomatic or have such mild symptoms that they do not seek medical care. These people are not part of the data set when determining death rates. Thus, he said the actual mortality rate "is way under 1%."[37]

The winner: President Trump. According to Caitlin Rivers, epidemiologist at the Johns Hopkins Center for Public Health, the current best estimate for fatality rates at the time were 0.5% to 1.0%.[38]

On the contrary, March 30, 2020, Fauci told CNN that there would be between 100,000 and 200,000 deaths from COVID-19 in the U.S.[39] Dr. Deborah Birx agreed, calling this "our real number." Based on these predictions, life as Americans knew it started to end. Lockdowns began and Americans began losing their rights, perhaps the most important of which was the right to assemble. All to prevent the spread of what was called one of the deadliest viruses of all time. Only it was not. On April 9, 2020, Fauci told Americans that the death toll would be more like 60,000.[40] This still sounds serious, although data for seasonal flu indicates that COVID-19 may be less deadly than we have been told.

Ginning up the numbers to gin up fear is a good way to get people to agree to vaccines for COVID-19. However, even vaccine advocate Paul Offit refused to engage in the deception. He stated publicly that the WHO's prediction of a 3.4% fatality rate was too high, and that the real number would likely be lower than 1.0%. "We're more the victim of fear than the virus," he said, and he also said that he thought the world was witnessing a "wild overreaction" to the disease.[41]

According to the Johns Hopkins Coronavirus Resource Center, on April 13, 2020 at 7:02AM there were 557,590 confirmed cases of COVID-19 and 22,109 deaths in the U.S.[42] The population of the U.S. is just over 330,000,000 which means that 0.17% of the population has been infected and that 3.97% of those infected had died. Johns Hopkins also reported that as more people are tested, more people with mild or asymptomatic cases will be identified, which will lower the death rate even more.[43]

In fact, this turned out to be the case when a Stanford University group randomly tested 3300 adults and children in Santa Clara County, CA. The group determined that the population prevalence of COVID-19 in Santa Clara was 2.49% - 4.16%, which means that most likely between 48,000 and 81,000 people had been infected in Santa Clara County – a 50- to 85-fold higher incidence rate than the number of infections predicted. As the denominator increased, the severity rate and death rate were shown to be much lower.[44]

Let us contrast this information with data from the Centers for Disease Control concerning seasonal flu. The CDC estimates that between October 1, 2019 and April 4, 2020 there were between 39 million and 56 million cases of flu; between 18 million and 26 million medical visits due to flu; between 410,000 and 740,000 hospitalizations due to seasonal flu; and between 24,000 and 62,000 deaths due to seasonal flu.[45]

These data reveal that COVID-19 appears to be significantly less dangerous than seasonal flu.

An analysis by Dr. John Ioannidis at Stanford University and colleagues confirmed this. The group analyzed data to determine the relative risk of death from COVID-19 in people younger than 65 years of age and those 65 and older. The analysis included data from Belgium, Germany, Italy, the Netherlands, Portugal, Spain, Sweden, Switzerland; Louisiana, Michigan, Washington states; and New York City. These areas included some places designated as "hot spots."

The analysis showed that:

- People under age 65 accounted for 5-9% of all COVID-19 deaths in 8 European countries
- People under age 65 had 34- to 73-fold lower risk than those 65 or older in European countries
- People under age 65 had 13- to 15-fold lower risk in NYC, Louisiana, and Michigan
- Absolute risk of COVID-19 death ranged from:
  1.7 per million for people under 65 in Germany (population incidence 0.0000017)
  to 79 per million in NYC (population incidence 0.000079)
- Absolute risk of COVID-19 death for people 80 or older ranged from 1 in 6000 in Germany to 1 in 420 in Spain
- The COVID-19 death risk in people <65 years old during the period of fatalities from the epidemic was equivalent to the death risk from driving between 9 miles per day (Germany) and 415 miles per day (New York City).
- People <65 years old and not having any underlying predisposing conditions accounted for only 0.3%, 0.7%, and 1.8% of all COVID-19 deaths in the Netherlands, Italy, and New York City.

"CONCLUSIONS: People <65 years old have very small risks of COVID-19 death even in the hotbeds of the pandemic and deaths for people <65 years without underlying predisposing conditions are remarkably uncommon."[46]

Many other people agreed that the response was completely disproportionate to the risk. Dr. Joel Kettner is professor of Community Health Sciences and Surgery for Manitoba University, former Chief Public Health Officer for Manitoba province, and Medical Director for the International Centre for Infectious Diseases. Here is what he said:

> "I have never seen anything like this, anything anywhere near like this. I'm not talking about the pandemic, because I've seen 30 of them, one every year. It is called influenza. And other respiratory illness viruses, we don't always know what they are. But I've never seen this reaction, and I'm trying to understand why.

> "I worry about the message to the public, about the fear of coming into contact with people, being in the same space as people, shaking their hands, having meetings with people. I worry about many, many consequences related to that."[47]

And there were many reasons to be concerned.

ENDNOTES

1. Pneumonia of Unknown Cause – China. World Health Organization 5 Jan 2020 https://www.who.int/csr/don/05-january-2020-pneumonia-of-unkown-cause-china/en/ accessed 9.1.2020

2. 41 confirmed cases of new coronavirus pneumonia Wuhan. *Xinhuanet* http://www.xinhuanet.com/2020-01/11/c_1125448269.htm accessed 9.1.200

3. Erin Schumaker. 1st confirmed case of new coronavirus reported in US: CDC. *ABC News* Jan 21 2020 https://abcnews.go.com/Health/1st-confirmed-case-coronavirus-reported-washington-state-cdc/story?id=68430795 accessed 9.1.2020

4. Lily Kuo. Coronavirus: panic and anger in Wuhan as China orders city into lockdown. *The Guardian* Jan 23 2020 https://www.theguardian.com/world/2020/jan/23/coronavirus-panic-and-anger-in-wuhan-as-china-orders-city-into-lockdown accessed 9.1.2020

5. Erin Schumaker. Coronavirus declared global health emergency by WHO after 1st person-to-person US case reported. *ABC News* Jan 30 2020 https://abcnews.go.com/Health/world-health-organization-decide-coronavirus-global-health-emergency/story?id=68639487 accessed 9.1.2020

6. Helen Regan, Jessie Yeung, Steve George, Amy Woodyat. January 29 coronavirus news. *CNN* Jan 29 2020 https://www.cnn.com/asia/live-news/coronavirus-outbreak-01-29-20-intl-hnk/index.html accessed 9.1.2020

7. Coronavirus: Epidemiological alerts and updates. Pan American Health Organization. 14 Feb 2020 https://www.paho.org/hq/index.php?option=com_docman&view=list&slug=coronavirus-epidemiological-alerts-and-updates&Itemid=270&layout=default&lang=en accessed 7.11.2020

8. Erin Schumaker. Coronavirus declared global health emergency by WHO after 1st person-to-person US case reported. *ABC News* Jan 30 2020 https://abcnews.go.com/Health/world-health-organization-decide-coronavirus-global-health-emergency/story?id=68639487 accessed 9.1.2020

9. Morgan Winsor, Erin Schumaker. 12th coronavirus case confirmed in the US. *ABC News* Feb 5 2020 https://abcnews.go.com/International/10-people-aboard-cruise-ship-japan-test-positive/story?id=68769789 accessed 9.1.2020

10. Tedros Adhanon Ghebreysus. Twitter. Feb 11 2020 https://twitter.com/DrTedros/status/1227297754499764230 accessed 9.1.2020

11. CDC, Washington State Report First COVID-19 Death. Centers for Disease Control. https://www.cdc.gov/media/releases/2020/s0229-COVID-19-first-death.html accessed 9.1.2020

12. Helen Branswell. WHO declares the coronavirus outbreak a pandemic. *STAT* Mar 11 2020 https://www.statnews.com/2020/03/11/who-declares-the-coronavirus-outbreak-a-pandemic/ accessed 9.1.2020

13. Libby Cathey. Government coronavirus response: Trump declares national emergency, says he 'likely; will get tested. *ABC News* March 13 2020 https://abcnews.go.com/Politics/government-coronavirus-response-live-updates-trump-declares-national/story?id=69580277 accessed 9.1.2020

14. Reuters. CDC Reports 1,678 Coronavirus Cases, Death Tally of 41. *US News* March 13 2020 https://www.usnews.com/news/us/articles/2020-03-13/us-cdc-reports-1-678-coronavirus-cases-death-tally-of-41 accessed 9.1.2020

15. Emily Shapiro, Jon Haworth, Karma Allen. Coronavirus shuts down major cities, Trump asks Americans to avoid groups over 10 people. *ABC News* March 17 2020 https://abcnews.go.com/US/

coronavirus-live-updates-establishments-country-begin-shutting/story?id=69615056 accessed 9.1.202

16. China reports no new local coronavirus transmissions for first time. *Reuters* March 18 2020 https://www.reuters.com/article/us-health-coronavirus-china/china-reports-no-new-local-coronavirus-transmissions-for-first-time-idUSKBN216085 accessed 9.1.2020

17. Morgan Winsor, Emily Shapiro, Ella Torres. NYC now epicenter of coronavirus, mayor says: 258 dead in US." *ABC News* Mar 20 2020 https://abcnews.go.com/Health/coronavirus-live-updates-china-exonerates-whistleblower-doctor-warned/story?id=69702910 accessed 9.1.2020

18. 43 Coronavirus deaths and Over 5600 Cases in N.Y.C. *New York Times* Mar 20 2020 https://www.nytimes.com/2020/03/20/nyregion/coronavirus-new-york-update.html accessed 9.1.2020

19. CDC COVID Data Tracker. Centers for Disease Control and Prevention https://www.cdc.gov/coronavirus/2019-ncov/cases-updates/us-cases-deaths.html accessed 7.11.2020

20. COVID-19 Coronavirus Tracker. Centers for Disease Control and Prevention https://www.kff.org/coronavirus-covid-19/fact-sheet/coronavirus-tracker/ accessed 7.11.2020

21. Lambert LC, Fauci AS. "Influenza vaccines for the future." *NEJM* 2010 Nov;363(21):2036-2044

22. Centers for Disease Control and Prevention. Estimates of deaths associated with seasonal influenza – United States, 1976-2007. *MMWR Morb Mortal Wkly Rep* 2010 Aug;59(33):1057-1062

23. Lambert LC, Fauci AS. "Influenza vaccines for the future." *NEJM* 2010 Nov;363(21):2036-2044

24. Centers for Disease Control and Prevention. "Estimates of deaths associated with seasonal influenza – United States, 1976-2007." *MMWR Morb Mortal Wkly Rep* 2010 Aug;59(33):1057-1062

25. Simonsen L, Clarke MJ, Williamson GD, Stroup DF, Arden NH, Schonberger LB. "The impact of influenza epidemics on mortality: introducing a severity index." *Am J Public Health* 1997 Dec;87(12):1944-1950

26. Simonsen L, Fukuda K, Schonberger LB, Cox NJ. "The impact of influenza epidemics on hospitalizations." *J Infect Dis* 2000 Mar;181(3):831-837

27. Thompson WW, Shay DK, Weintraub W et al. "Influenza-associated hospitalizations in the United States." *JAMA* 2004 Sep;292(11):1333-1340

28. Molinari NAM, Ortega-Sanchez IR, Messonnier ML et al. "The annual impact of seasonal influenza in the US: measuring disease burden and costs." *Vaccine* 2007 Jun;25(27):5086-5096

29. Ferguson NM, Laydon D, Nedjati-Gilani G et al. Report 9: Impact of non-pharmaceutical nterventions (NPIs) to reduce COVI-19 mortality and healthcare demand." *Imperial College COVID-19 Response Team* March 16 2020

30. Imperial College of London News Release. Swine flu: early findings about pandemic potential reported in a new study. Imperial college of London May 12 2009 https://www.imperial.ac.uk/news/66374/swine-early-findings-about-pandemic-potential/ accessed 9.1.2020

31. Six questions that Neil Ferguson should be asked. *The Spectator* April 16 2020 https://www.spectator.co.uk/article/six-questions-that-neil-ferguson-should-be-asked accessed 9.1.2020

32. James Sturcke. Bird flu pandemic could kill 150,000. *The Guardian* Sept 30 2005 https://www.theguardian.com/world/2005/sep/30/birdflu.jamessturcke#:~:text=%22The%20consequences%20in%20terms%20of,between%20five%20and%20150%20million%22. Accessed 9.1.2020

33. Fauci AS, Lane HC, Redfield RR. "." *NEJM* 2020 Mar;382:1268-1269

34. Ronald Bailey. "COVID-19 Mortality Rate 'Ten Times Worse' Than Seasonal Flu, Says Dr. Anthony Fauci." *Reason* Mar 11 2020 https://reason.com/2020/03/11/covid-19-mortality-rate-ten-times-worse-than-seasonal-flu-says-dr-anthony-fauci/ accessed 9.1.2020

35. Joseph Guzman. Coronavirus 10 times more lethal than seasonal flu, top health official says." *The Hill* Mar 11 2020 https://thehill.com/changing-america/well-being/prevention-cures/487086-coronavirus-10-times-more-lethal-than-seasonal accessed 9.1.2020

36. WHO Director-General's opening remarks at the media briefing on COVID-19 – March 2020. World Health Organization. https://www.who.int/dg/speeches/detail/who-director-general-s-opening-remarks-at-the-media-briefing-on-covid-19---3-march-2020 accessed 9.1.2020

37. Ibid

38. Jon Hamilton. Antibody Tests Point To Lower Death Rate For The Coronavirus Than First Thought. *WABE* May 28 2020 https://www.wabe.org/antibody-tests-point-to-lower-death-rate-for-the-coronavirus-than-first-thought/ accessed 9.1.2020

39. Devan Cole, Kevin Bohn, Dana Bash. US could see millions of coronavirus cases and 100,000 or more deaths, Fauci says. *CNN* March 30, 2010 https://www.cnn.com/2020/03/29/politics/coronavirus-deaths-cases-anthony-fauci-cnntv/index.html accessed 9.1.2020

40. Ian Schwartz. Dr. Fauci: Virus Death Toll May Be "More Like 60,000 Than 100,000 to 200,000. *Real Clear Politics* April 9 2020 https://www.realclearpolitics.com/video/2020/04/09/dr_fauci_virus_death_toll_may_be_more_like_60000_than_100000_to_200000.html accessed 98.1.2020

41. Rem Rider. Trump and the Coronavirus Death Rate. *Factcheck Posts* March 5 2020

42. https://coronavirus.jhu.edu/map.html

43. COVID-19 Dashboard by the Center for Systems and Engineering (CSSE) at Johns Hopkins University. Coronavirus Resource Center https://coronavirus.jhu.edu/data/mortality accessed 4.23.2020

44. Alexandra Hein. Coronavirus antibody testing finds Bay Area infections may be 85 times higher than reported: researchers. *Fox News* April 17 2020 https://www.foxnews.com/health/coronavirus-anti-body-testing-finds-bay-area-infections-85-times-higher-reported-researchers accessed 4.20.2020

45. 2019-2020 U.S. Flu Season: Preliminary Burden Estimates. Centers for Disease Control and Prevention. https://www.cdc.gov/flu/about/burden/preliminary-in-season-estimates.htm accessed 9.1.2020

46. Ioannidis JPA, Axfors C, Contopoulos-Ioannidis DG. "Population-level COVID mortality risk for non-elderly individuals overall and for non-elderly individuals without underlying diseases in pandemic epicenters." *Environ Res* 2020 Sep;188:109890

47. OffGuardian. 12 Experts Questioning the Coronavirus Panic. *Global Research: Center for Research on Globalization.* https://www.globalresearch.ca/12-experts-questioning-coronavirus-panic/5707532 accessed 9.1.2020

THE ORIGIN OF SARS-COV-2

WAS COVID-19 the virus that was the subject of gain-of-function research conducted by American researchers in conjunction with the Wuhan Institute of Virology? We may never know the answer to this question since the Chinese Communist Party (CCP) has not been willing to share information with the rest of the world. Even so, some background on the Wuhan Institute, the staff, and how the Chinese government responded provides some clues as to what may have happened.

The Wuhan Institute of Virology (WIV) was originally founded in 1956 as the Wuhan Microbiology Laboratory. The Institute has operated under the jurisdiction of the Chinese Academy of Sciences since 1978. The Institute's labs range from Biosafety Level II (BSL-2) to Biosafety Level IV (BSL-4). BSL-4 labs can be used for research with dangerous agents and substances.

The WIV BSL-4 LAB, which is of interest in the COVID-19 debacle, was developed by the People's Republic of China (PRC) in partnership with France following the 2003 SARS pandemic.

Almost immediately after the project was undertaken, French officials expressed discomfort because it was suspected that the PRC had a biological warfare program and the BSL-4 lab might be used for the purpose of developing biological weapons. To mitigate this concern, the parties agreed that all PRC/French research projects would be conducted under the supervision of French researchers on site at the lab. This did not, however, resolve the issue.

Disagreements between the parties continued. The French obtained information that led them to think that the PRC intended to build several BSL-4 labs. There were ongoing disputes over construction. After the lab opened, the French became alarmed when the PRC requested biohazard suits that offered protection beyond what would have been necessary based on the research that should have been going on in the lab.

Of concern to everyone is the influence the Chinese Communist Party (CCP) had and continues to have on the Institute. High-level CCP officials serve on committees that decide the projects that will be undertaken in the lab and are also placed in management positions.

Accidents at the lab have been another concern. For example, during a one-month period in 2004, the PRC reported nine new cases of SARS related to an accident during research using both live and inactivated samples of SARS-CoV.[1]

The Institute is headed by Dr. Shi Zheng-Li, who is known as China's "Bat Woman" because she has spent a significant portion of her career collecting bat viruses to make vaccines.[2] Her colleagues include scientists and physicians who have close ties to both the political and military leadership of the PRC. An example is Guo Deyin, who has conducted research on AIDS and hepatitis vaccines, as well as genetic recombination methods.

Dr. Shi's Research at WIV

In a 2010 paper, Shi and her colleagues reported the results of their research on angiotensin-converting enzyme II (ACE2) protein, which is a known SARS-CoV receptor. The group looked at ACE2 molecules from seven bat species and tested the interaction of the ACE2 receptor with the human SARS-CoV spike protein. They used HIV-based pseudo type and live SARS-CoV infection assays. Spike proteins are structures that allow coronaviruses to bind to the receptor sites on human cells.

The researchers found that the ACE2s of two bat species – *Myotis daubentoni* and *Rhinolophus sinicus* were susceptible to SARS-CoV and might be candidates as the natural host of the SARS-CoV progenitor viruses.[3]

Shi was also a member of the Chinese research team that was involved in the controversial gain-of-function research financed by the U.S. government, and conducted in partnership with a research team at the University of North Carolina Chapel Hill. In a paper published in 2015 in *Nature Medicine* the group characterized a chimeric virus with the spike protein SHC014 that was able to use multiple genes of the SARS receptor human angiotensin converting enzyme II (ACE2) and "replicate efficiently in primary human airway cells and achieve in vitro titers equivalent to epidemic strains of SARS-Cov." In other words, this virus could infect humans and quickly replicate. The article specifically stated, "...we synthetically re-derived an infectious full-length SHC014 recombinant virus and demonstrate robust viral replication both *in vitro* and *in vivo*."

Furthermore, the team also reported replication of the chimeric virus in the lungs of mice. Most important, therapies typically used to treat SARS patients were found to be ineffective for treating the chimeric virus and vaccines did not prevent "infection with CoVs using the novel spike protein."[4]

The bottom line: Researchers at the Wuhan lab were conducting research on bat viruses, were successful on at least one occasion in developing one that could infect humans, and this virus seemed to be resistant to treatment and prevention with vaccines.

The Outbreak Begins in China

There are significant discrepancies concerning the timeline and the actual events surrounding the outbreak. Stories changed, information was withheld, and important evidence was destroyed. For a long time, the World Health Organization reported almost verbatim information provided by the Chinese government, and based its recommendations based on that data. Even after it was discovered that the government had incorrectly (and it appears deliberately) attributed the outbreak to the Huanan Seafood Market, WHO made no effort to intervene or to discipline China for its misrepresentation.

Following is the timeline for events and how information became available in China, along with the WHO response:
The virus was first reported in Wuhan in Hubei Province, the largest city in Central China with a population of 11.2 million people in December 2019. There are varying accounts from the CCP concerning both when the first case was reported and when the CCP knew about it.

The first official announcement was issued on **December 30, 2019** when the Wuhan Municipal Health Commission reported that "cases of pneumonia of unknown cause" were linked to the Huanan Seafood Market, which sold live wild animals in addition to seafood, including hedgehogs, badgers, snakes, and turtledoves. It was also stated there was no evidence of "obvious human to human transmission and no infection among medical personnel."[5]

Also, on December 30, 2019, the Wuhan Municipal Health Commission issued an "urgent notice" to medical institutions ordering them to track and report cases right away to various district CDC's and to the Wuhan Municipal Health Commission.

But it appears that late December was not the actual beginning of the outbreak. In March 2020, the Hong Kong-based *South China Morning Post* reported that according to the Chinese government, the first known patient was a 55-year-old from Hubei who became ill on November 17, 2019.[6]

According to another article published in the *Lancet* on January 20, 2020, doctors from a Wuhan hospital reported that of the first 41 cases later identified as having COVID-19, the first one of those had symptoms on December 1, 2019.[7] Based on the incubation period, this person most likely was infected in November.

Gao Fu, Director of the China Centers for Disease Control, denied this, stating that "There is no solid evidence to say we already had clusters in November."[8] Chinese authorities later stated that the first known patient experienced symptoms on December 8, 2019.[9]

On January 1, 2020, the Huanan Seafood Market was closed for cleaning. Vendors reported that workers had started spraying disinfectant on December 30, 2019.[10]

Scientists from China's National Institute for Viral Disease Control and Prevention collected 515 samples from the Huanan Seafood Market for analysis on January 1, 2020 and returned to collect 70 more samples from vendors after the market re-opened.

At the same time, an official at the Hubei Provincial Health Commission ordered gene sequencing companies and labs to stop testing and to destroy all patient samples.[11]

On January 2, 2020 an analysis of samples from patients at Wuhan's Jinyintan Hospital by researchers at Wuhan Institute of Virology identified the novel coronavirus.[12]

On January 3, 2020 the Wuhan Municipal Commission reported that 44 patients had been identified with symptoms consistent with "pneumonia of unknown origin" some of whom worked at the Huanan Seafood Wholesale Market and 11 of whom were severely ill.[13]

January 5, 2020

The Wuhan Municipal Health Commission announced that it had identified 59 patients with symptoms consistent with "pneumonia of unknown origin." It stated that a preliminary investigation had uncovered no "clear evidence of human-to-human transmission" or infections among medical workers.[14]

Shi and her team asked the World Health Organization to register SARS-CoV-2 as a new virus, H-nCoV-19, rather than another virus derived from SARS.[15]

January 5, 2020

This statement was posted by WHO:

> On 31 December 2019, the WHO China Country Office was informed of cases of pneumonia of unknown etiology (unknown cause) detected in Wuhan City, Hubei Province of China. As of 3 January 2020, a total of 44 patients with pneumonia of unknown etiology have been reported to WHO by the national authorities in China. Of the 44 cases reported, 11 are severely ill, while the remaining 33 patients are in stable condition. According to media reports, the concerned market in Wuhan was closed on 1 January 2020 for environmental sanitation and disinfection.
>
> The causal agent has not yet been identified or confirmed. On 1 January 2020, WHO requested further information from national authorities to assess the risk.
>
> National authorities report that all patients are isolated and receiving treatment in Wuhan medical institutions. The clinical signs and symptoms are mainly fever, with a few patients having difficulty in

breathing, and chest radiographs showing invasive lesions of both lungs.

According to the authorities, some patients were operating dealers or vendors in the Huanan Seafood market. Based on the preliminary information from the Chinese investigation team, no evidence of significant human-to-human transmission and no health care worker infections have been reported.[16]

The WHO also posted this statement:

The reported link to a wholesale fish and live animal market could indicate an exposure link to animals. The symptoms reported among the patients are common to several respiratory diseases, and pneumonia is common in the winter season; however, the occurrence of 44 cases of pneumonia requiring hospitalization clustered in space and time should be handled prudently.[17]

In other words, the WHO was repeating the Chinese claim that the virus originated in the seafood market and gave the impression that there was no reason for concern.

January 7, 2020

A team led by Professor Yong-Zhen Zhang of Fudan University in Shanghai identified a novel coronavirus and sequenced its genome. The team reported its work to Chinese authorities and submitted the sequence to GenBank, a genetic sequence database operated by the U.S. National Institutes of Health that serves as "an annotated collection of all publicly available DNA sequences." The team also submitted an article to the journal *Nature* detailing the team's sequencing of the novel coronavirus.[18]

According to a March 26, 2020 paper published in the *New England Journal of Medicine,* China CDC also completed genomic sequencing of the novel coronavirus on January 7, 2020.[19]

China's National Health Commission issued a directive on management of biological samples in major infectious disease outbreaks. The directive "ordered institutions not to publish any information related to the unknown disease, and ordered labs to transfer any samples they had to designated testing institutions, or to destroy them."[20]

January 9, 2020

WHO issued this statement: "WHO does not recommend any specific measures for travelers. WHO advises against the application of any travel or trade restrictions on China based on the information currently available."[21]

January 10, 2020

WHO issues "Advice for International Travel and Trade in Relation to the Outbreak of Pneumonia Caused by a New Coronavirus in China." The agency recommended against entry screening for travelers, stating, "It is generally considered that entry screening offers little benefit, while requiring considerable resources."

Simply repeating information provided from the Chinese government, WHO states, "From the currently available information, preliminary investigation suggests that there is no significant human-to-human transmission, and no infections among health care workers have occurred."[22]

January 11, 2020

The Wuhan Municipal Health Commission announced the first death of a coronavirus patient, a 61-year-old man who was a long-time customer of the Huanan Seafood Wholesale Market. The commission states again that it had not found evidence of person-to-person transmission or infections among health care workers.[23]

WHO tweets, "BREAKING: WHO has received the genetic sequences for the novel #coronavirus (2019-nCoV) from the Chinese authorities. We expect them to be made publicly available as soon as possible."[24]

China later said the Chinese institutions that jointly shared the genomic sequence with WHO are China CDC, the Chinese Academy of Medical Sciences, and the Wuhan Institute of Virology under the Chinese Academy of Sciences, as designated agencies of the National Health Commission.[25]

But remember, on January 7, 2020 Chinese officials ordered samples to be sent to specific institutions or destroyed.

January 12, 2020

WHO issued this statement: "China shared the genetic sequence of the novel coronavirus on 12 January, which will be of great importance for other countries to use in developing specific diagnostic tests." WHO also stated, "The evidence is highly suggestive that the outbreak is associated with exposures in one seafood market in Wuhan. The market was closed on 1 January 2020. At this stage, there is no infection among healthcare workers, and no clear evidence of human to human transmission."[26]

January 14, 2020

WHO tweets, "Preliminary investigations conducted by the Chinese authorities have found no clear evidence of human-to-human transmission of the novel #coronavirus (2019-nCov) identified in #Wuhan, #China."[27]

January 26, 2020

The Institute of Virology and Chinese CDC announced that the novel coronavirus was present in 33 of the 585 environmental samples collected from the Wuhan Huanan Seafood Wholesale Market earlier in the month. Of these 33 samples, all but two were collected from an area of the market where wildlife vendors were located. Xinhua News Service says the results indicate "the virus stems from wild animals on sale at the market."[28]

Almost immediately, however, published research showed that the market could not have been the source of the outbreak. The co-authors of an article published in the *Lancet*, including experts from Wuhan's leading infectious disease hospital, reported that among the first 41 patients identified in Wuhan, the first patient to show symptoms, on December 1, 2019, had no exposure to the market. Two of the next three patients to show symptoms, all on December 10, also had no exposure to the market. "No epidemiological link was found between the first patient and later cases," the researchers wrote. And, in fact, there were 13 patients with no link to the market.[29]

"That's a big number, 13, with no link," stated Daniel Lucey, an infectious disease specialist at Georgetown University, who went on to say that the *Lancet* paper raised questions about the overall accuracy of the data the CCP was providing to the world.

According to Lucey, the Wuhan Municipal Health Commission was the "official source" of public information and on January 11, 2020 reported that that there were only 41 confirmed patients, that there was no evidence of human-to-human transmission, and that most cases were related to the market. Because the Wuhan Municipal Health Commission noted that diagnostic tests had confirmed these 41 cases by January 10, 2020 and officials presumably knew the case histories of each patient, Lucey said "China must have realized the epidemic did not originate in that Wuhan Huanan seafood market."[30]

Kristian Andersen, an evolutionary biologist at the Scripps Research Institute, analyzed sequences of 2019-nCoV to try to clarify its origin. Andersen posted an analysis of 27 available genomes of COVID-19 on a virology website and suggested they had a "most recent common ancestor"—meaning a common source—as early as 1 October 2019.[31]

An article published in the *Lancet* on January 30, 2020 reported that of 99 patients diagnosed with COVID-19 between Jan 1 and Jan 20, 2020, forty-nine had been exposed to the Huanan Seafood Market, and 50 had not.[32] And an article in the *New England Journal of Medicine* reported that of 425 confirmed cases, the majority (55%) with onset before January 1, 2020 were linked to seafood market, although this was true for only 8.6% of subsequent cases.[33] The theory that the seafood market was the source of the outbreak and that the virus was not transmissible between humans was falling apart.

It is important to note that the First National Health Commission arrived in Wuhan December 31, 2019 and determined that in order to diagnose SARS-CoV-2, three criteria needed to be met: a history of exposure to the seafood market, fever, and the full genome from respiratory or serum specimens identical to SARS-CoV-2 sequences.[34]

The timeline above, however, indicates that the Chinese knew that one third had no contact with the seafood market when these criteria were established. So why were these criteria established? To mislead the world about the origin of the virus? The criteria were not changed until January 18, 2020 but on January 26, 2020 Chinese authorities were still claiming that the virus originated at the seafood market.

So where *did* this virus originate?

A sample of bronchoalveolar fluid from a single patient hospitalized on December 26, 2019 identified a new RNA virus strain most closely related (89.1% nucleotide similarity) to a group of SARS-like coronaviruses previously found in bats in China. The researchers noted that although SARS-like viruses have been identified widely in bats in China, viruses identical to SARS-CoV had not yet been documented. They noted that the Wuhan coronavirus was most closely related to bat coronaviruses, and showed 100% amino acid similarity to bat SL-CoVZC45 in the nsp7 and E proteins. [35] The problem is that there were no bats at the seafood market, which means that the virus could not have originated there.

In a paper published in the *Lancet,* researchers wrote, "Notably, 2019-nCoV was closely related (with 88% identity) to two bat-derived severe acute respiratory syndrome (SARS)-like coronaviruses, bat-SL-CoVZC45 and bat-SL-CoVZXC21, collected in 2018 in Zhoushan, in eastern China."[36] The researchers were referring to a 2018 paper which reported the results of an analysis of 334 bats collected between 2015 and 2017 from Zhoushan City in Zhejiang province China. Coronaviruses were detected in 26.65% of these bats, and the viruses had 81% shared nucleotide identity with human/civet SARSCoVs.[37] This sounds complicated and it is, but what this means is that the Wuhan virus was very similar to bat viruses. Yet there were no bats at the seafood market. Also remember that "the bat lady" – Shi - had been studying bat viruses at the WIV for an exceptionally long time.

Again, the CCP was not forthcoming. The Shanghai lab where researchers published the first genome sequence of the coronavirus that caused COVID-19 was shut down by the Shanghai Health Commission for "rectification" on January 12, 2020, five days after Professor Yong-Zhen Zhang's team published the genome sequence and made it available to the public. The team had reported that the virus resembled a group of viruses previously found in bats. This lab was a Level 3 biosafety facility and had just passed its annual inspection on January 5, 2020.[38]

Indian researchers also studied the virus and found four insertions in the spike protein that are unique to SARS-CoV-2 and not present in other coronaviruses. The amino acid residues in all four insertions were found to be similar to amino acid residues in the structural proteins of HIV-1. The researchers noted that there are only 3 viruses that contain these sequences – HIV-1, the bat coronaviruses discovered by Shi, and the New Wuhan virus (COVID-19). They also noted that it was highly unlikely that this could have occurred naturally.[39]

This article was later withdrawn without comment. And notes from a lecture delivered by Shi shortly before the outbreak began disappeared from the Institute website.

The CCP's order to labs to destroy samples, and its refusal to share information and samples to the world community has not helped to instill confidence in the integrity of Chinese officials and their representations concerning the virus.[40]

More Interesting Info About Chinese Institutions

On January 21, 2020, The Wuhan Institute of Virology applied for a Chinese patent on Gilead's Remdesivir.[41]

Jiang Mianheng is the eldest son of former CCP leader Jiang Zemin.[42] Mianheng created the Shanghai Institute of Life Sciences, along with several medical facilities including hospitals in Shanghai and hospitals for the military. He has served as Vice President of the Chinese Academy of Sciences.[43] Jiang Zhicheng is Mianheng's son and owns controlling interest in Wuki AppTec, which controls Fosun Pharma,[44] China's agent for remdesivir. Fosun Pharma partnered with BioNTech to develop and introduce an nRNA vaccine for COVID-19.[45]

Five whistleblowers who reported irregularities at the Wuhan lab are missing and one is dead. Dr. Li Wenliang tried to warn the world about the virus. The police sent him a letter shortly before his death warning that "if he refused to repent he would be punished." He reportedly died of COVID-19 at the age of 34.

- Ren Zhiqiang openly criticized President Xi for the CCP's handling of the virus. He has been missing since March.
- Chen Quishi, a Chinese citizen journalist, went missing in February after exposing the severity of the virus. His family was told he was in medical quarantine and he has not been seen since.

- Fang Bin's laptop was seized by police after taking pictures of eight dead bodies outside a Wuhan hospital. He vanished on February 8, 2020.
- Li Zehua is a 25-year-old journalist who visited the Wuhan lab on Feb 26, 2020 and has not been seen since.
- Xu Zhangrun was placed under house arrest in Beijing after criticizing President Xi. His internet service has been disconnected and he has been barred from social media.[46]
- Institute researcher Chen Quanjiao blew the whistle reporting that Director General of the Institute Wang Yanyi was suspected of leaking the virus.[47] Chen later retracted the accusation.[48]

It seems that people who dare to speak up about how COVID-19 is handled in China do not fare well.

The World Health Organization

The World Health Organization's response to COVID-19 was not in accordance with its own guidelines. WHO officials ignored a warning from the University of Hong Kong Center of Infection issued on January 4, 2020. The UHK School of Public Health has been a WHO Collaborating Centre for Infectious Disease Epidemiology and Control since 2014. Dr. Ho Chung Man notified WHO that based on the number of cases, it was likely that human-to-human transmission had already begun.

WHO guidelines also require that WHO notify member states as soon as possible even when unofficial reports concerning infectious disease outbreaks are received.

Director-General Tedros should have known by January 23, 2020 that the seafood market was not the source of the outbreak; that human-to-human transmission was taking place; that healthcare workers were being infected; and that at least four other countries had reported cases in addition to Hong Kong and Taiwan. He declined to declare a Public Health Emergency of International Concern (PHEIC), and instead visited Beijing. While there he praised the CCP's handling of the virus, and praised the CCP for its "transparency" in sharing information with the WHO and the rest of the world. Seven days later he declared a PHEIC. People have hypothesized that the reason for the delay was because of Tedros' relationship with Xi, and the fact that Xi's wife is on the goodwill council of WHO.

Even after the U.S. instituted travel restrictions on January 31, 2020, Tedros continued to insist that this was not necessary and would "…interfere with international travel and trade."[49]

Unanswered Questions

At this time, there is no way to know if the outbreak started at the WIV, or the actual origin of the virus.

Both the CCP and the WHO owe the world an explanation for the decisions and actions taken during the last several months. The CCP needs to explain the following to the world:

- Why the Wuhan lab staff were not the first responders. It seems that these people would have been the most capable of evaluating the situation and formulating a plan.
  - Why reports were not made to the WHO in a timely manner
  - Why outside experts were not permitted to visit the Wuhan Institute of Virology
  - Why officials closed the Shanghai lab where genomic sequencing was performed and sealed off all access to international researchers
  - Why labs were told to destroy their samples
  - Why officials continued to attribute the outbreak to the Huanan market long after it was apparent that this was not the case
  - Why the Wuhan lab was not searched
  - If this virus was the result of gain-of-function research conducted at the WIV or other sites
  - The status and location of the whistleblowers
  - Why there is such secrecy concerning the virus

As for Tedros and the WHO, the world needs to know:

- Why the CCP was not sanctioned for failure to notify WHO about the outbreak in a timely manner
  - Why the WHO has not asked the above questions of the CCP
  - Why member countries were not notified about SARS-CoV-2 as required

- Whether or not viral samples and sequences have been obtained from the PRC and if not, why not?

As you will later learn, Tedros also needs to explain why he declared SARS-CoV-2 a pandemic when the number of cases did not indicate that it was, and why he has failed to withdraw the pandemic label even today when the cases and deaths do not justify the designation. Perhaps most importantly, Tedros should be asked why he and the WHO – the organization that is supposed to protect the health of the world's seven billion people - stood by while most countries adopted disastrous and draconian policies that harmed more people than were helped.

ENDNOTES

1. The Origins of the COVID-19 Global Pandemic, Including the Roles of the Chinese Communist Party and the World Health Organization. House Foreign Affairs Committee Minority Staff Interim Report. June 12. 2020 https://gop-foreignaffairs.house.gov/wp-content/uploads/2020/08/Interim-Minority-Report-on-the-Origins-of-the-COVID-19-Global-Pandemic-Including-the-Roles-of-the-CCP-and-WHO-8.17.20.pdf accessed 9.1.2020

2. Jane Qiu "How China's 'Bat Woman' Hunted Down Viruses from SARS to the New Coronavirus." *Scientific American* June 1 2020

3. Hou Y, Peng C, Yu M et al. "Angiotensin-converting enzyme 2 (ACE2) proteins of different bat species confer variable susceptibility to SATS-CoV entry." *Arch Virol* 2010;155(10):1563-1569

4. Menachery VD, Yount BL, Debbink K et al. "A SARS-like cluster of circulating bat coronaviruses shows potential for human emergence." *Nat Med* 2015 Nov;21:1508-1513

5. Zhang Jingshu and Wang Ruiwen Editor: Li Jie. Wuhan Central Hospital claims that SARS rumors spread through the internet, there is no doubt that the patient may be diagnosed. *Beijing News* 12.31.2019 http://www.bjnews.com.cn/news/2019/12/31/668421.html accessed 9.1.2020

6. Josephine Ma. Coronavirus: China's First Confirmed Covid-19 Case Traced Back to November 17. *South China Morning Post*, March 13 2020, https://www.scmp.com/news/china/society/article/3074991/coronavirus-chinas-first-confirmed-covid-19-case-traced-back. Accessed 8.11.2020

7. Huang C, Wang Y, Li X et al. "Clinical Features of Patients Infected with 2019 Novel Coronavirus in Wuhan, China." *Lancet,* 2020 Feb;395(10223):P497-506

8. Cohen J. "Not Wearing Masks to Protect Against Coronavirus Is a 'Big Mistake,' Top Chinese Scientist Says." *Science*, March 27, 2020, https://www.sciencemag.org/news/2020/03/not-wearing-masks-protect-against-coronavirus-big-mistake-top-chinese-scientist-says. Accessed 9.1.2020

9. Zhang Y. The Novel Coronavirus Pneumonia Emergency Response Epidemiology Team. "The Epidemiological Characteristics of an Outbreak of 2019 Novel Coronavirus Diseases (COVID-19)— China, 2020," *China CDC Weekly*, 2020 Feb;2(8): 113-122.

10. Seafood market closed after outbreak of 'unidentified' pneumonia. *Global Times* Jan 1 2020 https://www.globaltimes.cn/content/1175369.shtml accessed 9.1.2020

11. The Origins of the COVID-19 Global Pandemic, Including the Roles of the Chinese Communist Party and the World Health Organization. House Foreign Affairs Committee Minority Staff Interim Report. June 12. 2020 https://gop-foreignaffairs.house.gov/wp-content/uploads/2020/08/Interim-Minority-Report-on-the-Origins-of-the-COVID-19-Global-Pandemic-Including-the-Roles-of-the-CCP-and-WHO-8.17.20.pdf accessed 9.1.2020

12. Report of the WHO-China Joint Mission on Coronavirus Disease 2019 (COVID-19) 16-24 Feb 2020 https://www.who.int/docs/default-source/coronaviruse/who-china-joint-mission-on-covid-19-final-report.pdf accessed 9.1.2020

13. Lu H, Stratton CW, Tang YW. "Outbreak of Pneumonia of Unknown Etiology in Wuhan China: The mystery and the miracle." *J Med Viro* 2020 Apr;92(4):401-402

14. Report of the WHO-China Joint Mission on Coronavirus Disease 2019 (COVID-19) 16-24 Feb 2020 https://www.who.int/docs/default-source/coronaviruse/who-china-joint-mission-on-covid-19-final-report.pdf accessed 9.1.2020

15. Jiang S, Shi Z, Shu Y et al. "A distinct name is needed for the new coronavirus." *Lancet* 2020 Mar;395(10228):949

16. World Health Organization. Pneumonia of unknown cause – China. World Health Organization https://www.who.int/csr/don/05-january-2020-pneumonia-of-unkown-cause-china/en/ accessed 8.11.2020

17. IBID

18. Wu F, Zhao S, Yu B et al. "A New Coronavirus Associated with Human Respiratory Disease in China." *Nature* 2020 Mar;579(7798):265-269

19. Li Q, Guan X, Wu P et al. "Early Transmission Dynamics in Wuhan, China, of Novel Coronavirus-Infected Pneumonia." *NEJM* 2020 Mar;382(13):1199-1207

20. Gao Yu, Peng Yanfeng, Yang Rui, et al., "In Depth: How Early Signs of a SARS-Like Virus Were Spotted, Spread, and Throttled." *Caixin Global* February 29, 2020,

21. World Health Organization. WHO Statement regarding cluster of pneumonia cases in Wuhan, China. Jan 9 2020 https://www.who.int/china/news/detail/09-01-2020-who-statement-regarding-cluster-of-pneumonia-cases-in-wuhan-china

22. Advice for International Travel and Trade in Relation to the Outbreak of Pneumonia Caused by a New Coronavirus in China. World Health Organization January 10, 2020 https://www.who.int/news-room/articles-detail/who-advice-for-international-travel-and-trade-in-relation-to-the-outbreak-of-pneumonia-caused-by-a-new-coronavirus-in-china accessed 9.1.2020

23. Andrew Joseph. "First death from Wuhan pneumonia outbreak reported as scientists release DNA sequence of virus." *STAT* Jan 11 2020 https://www.statnews.com/2020/01/11/first-death-from-wuhan-pneumonia-outbreak-reported-as-scientists-release-dna-sequence-of-virus/ accessed 9.1.2020

24. World Health Organization https://twitter.com/WHO/status/1216108498188230657 accessed 9.1.2020

25. China publishes timeline on COVID-19 information sharing, cooperation. *Xinhuanet* Apr 6 2020 http://www.xinhuanet.com/english/2020-04/06/c_138951662.htm accessed 9.1.2020

26. Novel Coronavirus—China. World Health Organization. January 12, 2020 https://www.who.int/csr/don/12-january-2020-novel-coronavirus-china/en/ accessed 9.1.2020

27. https://twitter.com/WHO/status/1217043229427761152

28. China Detects Large Quantity of Novel Coronavirus at Wuhan Seafood Market. *XinhuaNet* January 27, 2020 http://www.xinhuanet.com/english/2020-01/27/c_138735677.htm accessed 9.1.2020

29. Huang C, Wang Y, Li X et al. "Clinical Features of Patients Infected with 2019 Novel Coronavirus in Wuhan, China." *Lancet,* 2020 Feb;395(10223):P497-506

30. Jon Cohen. Wuhan seafood market may not be source of novel virus spreading globally. *Science* Jan 26 2020 https://www.sciencemag.org/news/2020/01/wuhan-seafood-market-may-not-be-source-novel-virus-spreading-globally accessed 9.1.2020

31. Clock and TMRCA based on 27 genomes. Novel 2019 coronavirus. *Genomic Epidemiology* https://virological.org/t/clock-and-tmrca-based-on-27-genomes/347

32. Chen N, Zhou M, Dong X, Qu J, Gong F, Han Y. "Epidemiological and clinical characteristics of 99 cases of 2019 novel coronavirus pneumonia in Wuhan, China: a descriptive study." *Lancet* 2020 Feb;395(10223):P507-513

33. Li Q, Med M, Guan X et al. "Early Transmission Dynamics in Wuhan, China, of Novel Coronavirus-Infected Pneumonia." *NEJM* 2020 Mar;382:1199-1207

34. Han Y, Yang H. "The transmission and diagnosis of 2019 novel coronavirus infection disease (COVID-19): A Chinese perspective." *J Med Virol* 2020 Mar;92:639-644

35. Wu F, Zhao S, Yu B et al. "A new coronavirus associated with human respiratory disease in China." *Nature* 2020 Feb;579:265-269

36. Lu R, Zhao X, Li J et al. "Genomic characterization and epidemiology of 2019 novel coronavirus: implications for virus origins and receptor binding." *Lancet* 2020 Feb;395:565-574

37. Hu D, Zhu C, Ai L et al. "Genomic characterization and infectivity of a novel SARS-like coronavirus in Chinese bats." *Emerg Microbes Infect* 2018 Sep;7:154

38. Zhuang Pinghui "Chinese laboratory that first shared coronavirus genome with world ordered to close for 'rectification', hindering its Covid-19 research." *South China Morning Post* Feb 28 2020 https://www.scmp.com/news/china/society/article/3052966/chinese-laboratory-first-shared-coronavirus-genome-world-ordered accessed 9.1.2020

39. Pradhan P, Pandley AK, Mishra A et al. "Uncanny similarity of unique inserts in the 2019-nCoV spike protein to HIV-1 gp120 and Gag." *BioRxiv* https://doi.org/10.1101/2020.01.30.927871

40. IBID

41. Wuhan Institute of Virology Applies for a Patent on Gilead's Remdesivir. *The National Law Review* Feb 6 2020 https://www.natlawreview.com/article/wuhan-institute-virology-applies-patent-gilead-s-remdesivir accessed 9.1.2020

42. Jim Liao "Dawning Information Industry and Jiang Mianheng have Close Ties." *Epoch Times* Jun 28 2019 https://epochtimes.today/dawning-information-industry-and-jiang-mianheng-have-close-ties/ accessed 9.1.2020

43. http://www.crunchbase.com/person/jiang-mianheng accessed 9.1.2020

44. Update: US NH Awarded Wuhan Lab that Studied Bat Coronavirus a $#.7 Million US Grant. April 16 2020 https://gsiexchange.com/update-us-nih-awarded-wuhan-lab-that-studied-bat-coronavirus-a-3-7-million-us-grant/ accessed 9.1.2020

45. Tuba Khan "BioNTech and Fosun Pharma form COVID-19 vaccine strategic alliance in China." *PharmaShots* August 13 2020 https://pharmashots.com/press-releases/biontech-and-fosun-pharma-form-covid-19-vaccine-strategic-alliance-in-china/ accessed 9.1.2020

46. Brittany Vonow. "Without a Trace. Five Wuhan whistleblowers still missing and one is dead after exposing true horrors of coronavirus." *The US. Sun* April 19 2020 https://www.the-sun.com/news/705139/wuhan-whistleblowers-missing-one-dead-coronavirus/ accessed 9.1.2020

47. Goyal VK, Sharma C. "The novel coronavirus 2019: A naturally occurring disaster or a biological weapon against humanity: A critical review of the tracing the origin of novel coronavirus 2019." *J Entomol Zool Studies* 2020;8(2):01-05

48. Wuhan virologist warns of overseas-started 'leaked virus' conspiracy. *Global Times* Feb 17 2020 https://www.globaltimes.cn/content/1179908.shtml accessed 9.1.2020

49. The Origins of the COVID-19 Global Pandemic, Including the Roles of the Chinese Communist Party and the World Health Organization. House Foreign Affairs Committee Minority Staff Interim Report. June 12. 2020 https://gop-foreignaffairs.house.gov/wp-content/uploads/2020/08/Interim-Minority-Report-on-the-Origins-of-the-COVID-19-Global-Pandemic-Including-the-Roles-of-the-CCP-and-WHO-8.17.20.pdf accessed 9.1.2020

THE LOCKDOWN

IN MARCH 2020, most Americans were told by their state governments that it was necessary to "shelter in place," which is really another term for "house arrest." Selective closure of businesses followed. Local and state governments established varying criteria for which businesses could remain open and which would close. Businesses labeled as "essential" included supermarkets and grocery stores, big box stores, drug stores, convenience and discount stores, healthcare facilities, daycare centers, hardware stores, gas stations, auto repair shops, banks, post offices, shipping offices and companies, vet clinics, pet stores, educational institutions to facilitate distance learning, transportation, and businesses that could justify staying open because they were required to facilitate other essential businesses – computer and office supply stores for example. Restaurants could continue to operate if they only offered food for take-out and delivery.

Everything else was closed – theaters, gyms, salons, museums, sporting events, concert venues, and so on. The determination of which businesses were considered "essential" versus "non-essential" was fairly capricious since many of the "essential" businesses like big box stores sold the same products as those deemed "non-essential." In "COVIDLAND" it was ok to purchase shoes at Walmart, but not at a smaller independent shoe store. It was ok to purchase decorative items at Target but not at a boutique. It was ok to purchase clothing at Meijer but not at a smaller shop. Tobacco stores, cannabis dispensaries, and pawn shops were essential, yet many businesses that families had spent 25 or more years building were not.

The reason for the lockdown was to slow or stop the spread of COVID-19 and preserve hospital capacity, since it was predicted that millions of Americans would contract COVID, millions would require hospitalization and two million people would die. Americans were told preserving hospital capacity was the goal. While many people were against the lockdowns (we were, and for the record our families had income from businesses deemed "essential"), most were willing to give the government the benefit of the doubt for two weeks.

The problem was that the lockdowns did not end in two weeks, and instead continued for months. At the time this book was being finished (August 2020), few places in the U.S. were completely "open," and some states were going backwards and locking down again. Most states had re-opened some businesses with limitations, but entire industries, like theatre and the arts and sports events had not returned. Many businesses could only operate with severe restrictions such as the number of people allowed inside at one time, or in the case of restaurants, using only 25%-40% of seating capacity.

In addition to house arrest and business closures, social distancing, masks, incessant hand washing, and other practices were mandated too. Was there evidence to support these measures?

The Origin of Lockdowns: A High School Project

The only thing more shocking than placing hundreds of millions of people on house arrest is the fact that the origin of the idea was a class project completed by a 14-year-old in Albuquerque, New Mexico. Laura Glass created a computer simulation of a bird flu pandemic in a virtual town of 10,000 people as part of an Intel International Science and Engineering Fair. She computed how family members, co-workers, students, and people in social situations interact, and determined that the average teenager was in close contact with about 140 people daily, which was the most of any group. She reported that high school students had the greatest potential to spread disease. She hypothesized that adults bring diseases into communities, and infect children, after which a disease spreads through schools.

In one of Laura's simulations about half of the population of 10,000 becomes infected but by closing the schools the number was reduced to 500.[1]

A foundational paper was written by Laura, her father and two others called "Targeted Social Distancing Designs for Pandemic Influenza (2006)."[2] The paper concluded, "Implementation of social distancing strategies is challenging. They likely must be imposed for the duration of the local epidemic and possibly until a strain-specific vaccine is developed and distributed. If compliance with the strategy is high over this period, an epidemic within a community can be averted. However, if neighboring communities do not also use these interventions, infected neighbors will continue to introduce influenza and prolong the local epidemic, albeit at a depressed level more easily accommodated by healthcare systems."

Robert J. Glass, Laura's father, had no medical training, and no expertise in immunology or epidemiology. And Laura was a high school student. So how in the world did their hypothesis become the basis for a series of decisions that brought the economy to a halt, resulted in the highest unemployment rates in the history of the U.S., disrupted education for tens of millions of children, and as yet uncalculated harm to millions of people and even death for some?

This is a long story that goes back to 2006 when President George W. Bush asked experts to submit potential plans for addressing a flu epidemic, should one occur. The avian flu had circulated in 2006 and was not declared a pandemic. Although it had caused government agencies to become concerned about planning for a pandemic should one occur. Two government doctors, Dr. Richard Hatchett and Dr. Carter Mecher were aware of this paper and proposed what we refer to today as "shelter at home" and lockdown as potential strategies.

At the time, the idea was not considered to be practical. Dr. D.A. Henderson, who had led the international effort to eradicate smallpox, thought it was a terrible idea. In a paper he authored with an infectious disease expert, an epidemiologist, and another physician, he wrote that a lockdown would "result in significant disruption of the social functioning of communities and result in possibly serious economic problems."[3]

The text of the article is worth reading (emphasis ours):

There are **no historical observations or scientific studies that support the confinement by quarantine of groups** of possibly infected people for extended periods in order to slow the spread of influenza. ... It is difficult to identify circumstances in the past half-century when large-scale quarantine has been effectively used in the control of any disease. The negative consequences of large-scale quarantine are so extreme (forced confinement of sick people with the well; complete restriction of movement of large populations; difficulty in getting critical supplies, medicines, and food to people inside the quarantine zone) that **this mitigation measure should be eliminated from serious consideration...**

Home quarantine also raises ethical questions. Implementation of home quarantine could result in healthy uninfected people being placed at risk of infection from sick household members. Practices to reduce the chance of transmission (hand-washing, maintaining a distance of 3 feet from infected people) could be recommended, but a policy imposing home quarantine would preclude, for example, sending healthy children to stay with relatives when a family member becomes ill. Such a policy would also be particularly hard on and dangerous to people living in close quarters, where the **risk of infection would be heightened....**

Travel restrictions, such as closing airports and screen-ing travelers at borders have historically been ineffective. The World Health Organization Writing Group concluded that "screening and quarantining entering travelers at internation-al borders did not substantially delay virus introduction in past pandemics . . . and will likely be even less effective in the mod-ern era.".... It is reasonable to assume that the economic costs of shutting down air or train travel would be very high, and **the societal costs involved in interrupting all air or train travel would be extreme...**

During seasonal influenza epidemics, public events with an expected large attendance have sometimes been cancelled or post-poned, the rationale being to decrease the number of contacts with those who might be contagious. **There are, however, no certain indications that these actions have had any definitive effect on the severity or duration of an epidemic. Were consideration to be given to doing this on a more extensive scale and for an extended period, questions immediately arise as to how many such events would be affected. There are many social gather-ings that involve close contacts among people, and this prohi-bition might include church services, athletic events, perhaps all meetings of more than 100 people.** It might mean closing the-aters, restaurants, malls, large stores, and bars. **Implementing such measures would have seriously disruptive consequences...**

Schools are often closed for 1–2 weeks early in the de-velopment of seasonal community outbreaks of influenza pri-marily because of high absentee rates, especially in elementary schools, and because of illness among teachers. This would seem reasonable on practical grounds. **However, to close schools for longer periods is not only impracticable but carries the possibility of a serious adverse outcome....**

Thus, cancelling or postponing large meetings would not be likely to have any significant effect on the development of the epidemic. While local concerns may result in the closure of particular events for logical reasons, a policy directing communitywide closure of public events seems inadvisable.

Quarantine. As experience shows, there is no basis for recommending quarantine either of groups or individuals. **The problems in implementing such measures are formidable, and secondary effects of absenteeism and community disruption as well as possible adverse consequences, such as loss of public trust in government and stigmatization of quarantined people and groups, are likely to be considerable....**

The conclusion: Experience has shown that communities faced with epidemics or other adverse events respond best and with the least anxiety when **the normal social functioning of the community is least disrupted.** Strong political and public health leadership to provide reassurance and to ensure that needed medical care services are provided are critical elements. If either is seen to be less than optimal, **a manageable epidemic could move toward catastrophe.**

In other words, two doctors, an epidemiologist, and an infectious disease expert warned that a lockdown was not only inadvisable but would have a catastrophic effect on the population.

A 2007 report from the Centers of Disease Control highlighted serious consequences if a lockdown was implemented.

The report included the results of a survey conducted by Harvard University to determine the willingness of adults to agree to community mitigation in the event of a pandemic, and the expected consequences. Almost 75% reported that they would be willing to "curtail various activities" of daily life for a month. Over 94% said they would stay at home and away from other people for 7-10 days if they had the flu. They were not asked about longer periods of confinement.

When asked about financial difficulties associated with missed work, 74% reported that they could miss 7-10 days of work without financial problems, but 25% said even this amount of time would cause a problem. Most, or 57%, reported that they would experience severe financial hardship if they missed work for one month, and 76% thought that three months away from work would be financially disastrous.

The CDC report noted other consequences to be considered, including that millions of children who relied on school meals might not have adequate food. The report advised that planning and implementation of pandemic mitigation would require participation from all segments of society.[4]

Fauci, state health directors, and governors repeated over and over again the importance of "following the science," and "listening to the experts." It became apparent early on that no one was listening to any credible experts and there was no scientific basis for any of the decisions that were made.

ENDNOTES

1. Virtual city used to study flu pandemic. *UPI* May 9 2006 https://medicalxpress.com/news/2006-05-virtual-city-flu-pandemic.html

2. Glass RJ, Glass LM, Beyeler WE, Min HJ. "Targeted Social Distancing Designs for Pandemic Influenza." *Emerg Infect Dis* 2006 Nov;12(11):1671-1681

3. Inglesby T, Nuzzo JB, O'Toole T, Henderson DA. "Disease Mitigation Measures in the Control of Pandemic Influenza." *Biosecur Bioterror* 2006 Nov;4(4):366-375

4. Interim Pre-pandemic Planning Guidance: Community Strategy for Pandemic Influenza Mitigation in the United States – Early, Targeted, Layered Use of Nonpharmaceutical Interventions. Canters for Disease Control and Prevention. https://www.cdc.gov/flu/pandemic-resources/pdf/community_mitigation-sm.pdf

LIFE IN "COVIDLAND"

EVEN IF THEY were not justified, at first the lock-downs were fairly straight-forward. Essential businesses were given permission to remain open. Non-essential businesses were to close. People were supposed to shelter-in-place and only leave their homes to get food, prescription drugs or for other activities deemed "necessary." Gatherings were forbidden; sports, theater, movies – all prohibited. The original orders were for 14-30 days, a period designed to "flatten the curve" so that hospitals would not be overwhelmed.

Governors and their health department sidekicks started holding daily press briefings during which largely made-up numbers were presented to the public. Mantras like "flatten the curve," "we must save lives," and "if we only save one life it is all worth it!" were repeated again and again. People were both frightened and compliant; however, a couple of weeks turned into a month, and then several months.

Almost immediately many people around the world seemed to know that something was not right. The numbers did not add up, the policies made no sense, and there was a growing disconnect between the death rate and increasingly more restrictive rules. These people were some of the first to speak up about the horrific collateral damage that resulted from lockdowns and health policies. They were regularly attacked by "believers" who bought into the fake pandemic, who believed everything they were told by health authorities and news reporters and refused to even look at data that refuted their beliefs. One of the initial major consequences of the hoax was family members who stopped speaking to one another and the termination of long-term friendships and community ties. How did this happen? It was planned!

Propaganda is defined as information that is designed to persuade people to accept an idea or to join a cause, usually by using biased material and appealing to people's emotions. Dictators and despots have used propaganda throughout history to control their "subjects." One of the great masters of propaganda was Adolph Hitler.

Here is an excerpt from Hitler's "Mein Kampf", written in 1926 (Volume one, chapter six):

"The receptivity of the great masses is very limited, their intelligence is small, but their power of forgetting is enormous. In consequence of these facts, all effective propaganda must be limited to a very few points and must harp on these in slogans until the last member of the public understands what you want him to understand by your slogan. As soon as you sacrifice this slogan and try to be many-sided, the effect will piddle away, for the crowd can neither digest nor retain the material offered. In this way the result is weakened and in the end entirely cancelled out."[1]

Hitler established the Ministry of Public Enlightenment and Propaganda almost immediately after his election because he believed that controlling information was as important as controlling the military and the economy. Joseph Goebbels was appointed as his director of propaganda. Goebbels also understood the importance of propaganda, writing in his diary, "No one can say your propaganda is too rough, too mean; these are not criteria by which it may be characterized. It ought not be decent nor ought it be gentle or soft or humble; it ought to lead to success." In other words, anything that needed to be said that led to brainwashing was ok."[2] Goebbels said, "Think of the press as a great keyboard on which the government can play."[3] He was one of the most influential men in the Nazi era.

All of Hitler's henchmen were on board and knew how to stay "on message." Hermann Goring wrote "But after all it is the leaders of a country who determine the policy and it is always a simple matter to drag the people along, whether it is a democracy or fascist dictatorship, or a parliament or a communist dictatorship. Voice or no voice, the people can always be brought to the bidding of the leaders. That is easy. All you have to do is tell them they are being attacked, and denounce the peace makers for lack of patriotism and exposing the country to danger. It works the same in any country."[4]

We saw implementation of this approach as the COVID-19 crisis began to unfold. Simple slogans were used to gin up panic and compliance. "Cases" were reported constantly, a number that does not really matter since tens of thousands of people contract the flu in the US every year, and hundreds of millions of people contract it worldwide. We do not report "cases" of flu every 15 minutes during every flu season. We really do not pay much attention to death counts either. With COVID-19, constant harping on cases, and how fast they were increasing, kept people feeling uneasy and almost certain that danger loomed. The growing number of cases became referred to as a "crisis" and "pandemic" that was often compared to the Spanish flu in 2018.

"Flatten the curve" was established as a civic duty, along with sheltering at home, and shutting down the economy in order to stop the spread of the "dangerous and deadly virus." Health officials often teared up when thanking compliant citizens daily for doing as they were told, telling the public "you are all saving lives."

Yet another disgusting and misleading slogan was "We're all in this together." This was repeated daily by people who were still employed with generous salaries and benefits; were sheltering at home in their large residences with all of their needs met, or were able to go to their office daily and lead fairly normal lives. It is hard to imagine an unemployed restaurant worker worried about paying rent and feeding his family feeling comradery with wealthy doctors, news anchors and movie stars who repeated this mantra daily.

Anyone who questioned anything, ranging from the CDC's instructions to falsify death certificates to the collateral damage in the form of increased suicides, overdoses, bankruptcies, home-lessness, food insecurity, increased violence, child abuse, and so on, was immediately labeled as insensitive, incompetent, and obviously having no respect for human life.

People Are Shockingly Easy to Convince of Almost Anything

Many people asked us, "How can intelligent people be convinced of something so untrue and remain this way for so long?" An interesting experiment conducted in a high school classroom in 1967 showed that adherence to simple instructions and increasing control can be achieved within only a few days.

Ron Jones was a 25-year-old social studies teacher in Palo Alto California in 1967. During a discussion of the Holocaust, students asked how the Nazis convinced German citizens to become complicit with Hitler's plans. In response, Jones announced an experiment. He began acting a little sterner than usual and announced a new set of rules for the classroom. While he only intended for the rules to last for one day, the next morning the students were sitting upright at their desks and announced in unison, "Good morning Mr. Jones."

So, Jones added more rules. Student were told to salute one another with a Nazi-style hand gesture, stand to ask questions, questions could only contain three words, and work on a project to "eliminate democracy."

He also promoted unity with slogans and banners that read "Strength Through Involvement," and "Strength Through Discipline." Large gatherings were prohibited – accounts vary, but only 2-3 students could be together at any one time. Students were told that if they complied with the rules, they would receive an "A" and if they tried to resist in any way, they would be given an "F."

The experiment expanded to require participation when not at school. Students who failed to salute another student when encountering the student anywhere could be reported and fined. Students were encouraged to turn one another in.

One student later reported in a radio interview that the students became fearful, and even students who used to be friends saw their relationships frayed and distrust developed.

When the students started recruiting other students who were not part of Jones' class, he knew it was time to end it, which he did at a gathering in the auditorium during which he showed a film about Nazism.[5]

Sound familiar? It should. Adults, many considered smart and educated by their (often former) friends, started immediately believing everything they were told by government

officials and the media. They dutifully followed instructions, regardless of how irrational and draconian they were. And many became snitches when the government issued instructions to turn in anyone who dared to disobey.

Los Angeles Mayor Eric Garcetti encouraged citizens to report anyone who refused to follow his COVID-19-related orders and announced that "business ambassadors" accompanied by police officers were confronting violators and charging them with crimes. He said, "You know the old expression about snitches, well in this case snitches get rewards," Garcetti said. "We want to thank you for turning folks in and making sure we are all safe."[6] Mayor Bill de Blasio in New York City issued the same instructions.[7]

It Could Not Have Been Done Without the Media

A massive disinformation campaign orchestrated by an obedient media, staffed with "journalists" who seemed to have no curiosity or interest in investigating what was going on, was sustained for months. The government power grab was dependent on maintaining an atmosphere of fear. The media dutifully helped by constantly reporting patently false information. They reported that everyone was equally at risk of dying of COVID-19, and that almost any action taken by the government was justified.

Articles and videos, often authored by or featuring prominent doctors and researchers at prestigious institutions, were taken down if the content deviated from the simple slogans uttered daily by government and the media. Only a few journalists worldwide investigated the claims made by these qualified individuals and reported what they said or allowed them to be guests on radio and television shows.

Meanwhile, the rich and powerful decided to exempt themselves from "the rules."

First, it is worth noting that government and health officials who constantly lectured people about the importance of an almost indefinite lockdown had jobs that were considered "essential." It was easy for them to order other people to stay home. They were not at risk of losing their jobs or their incomes and their employers, the government, were not at risk of going out of business. This did not go unnoticed by thinking people. What was truly despicable, however, was the fact that they openly flouted the rules they imposed on others.

Neil Ferguson is the person who created the now-famous Imperial College of London model that turned out to be wrong (96% off!). He was one of the architects of the stay-at-home strategy that resulted in billions of people all over the world being locked down. Apparently, he did not think he was required to adhere to the rules he helped to create. He violated the order by inviting his married lover to visit him at his home in London on at least two occasions. He subsequently resigned his advisory position when he was caught. He acknowledged that this was a mistake and that he made an "error in judgment."

Scotland's chief medical officer, Catherine Calderwood, resigned after being caught near her family's second home in a different part of the country than her main address. She was issued a formal police warning because she left her main residence without a good reason.[8]

Emperor Jay Pritzker banned Amy Jacobson from attending press briefings, stating that she was "not being impartial." Jacobson was co-host of a morning drive time talk show on AM 560 with Dan Proft.

She was the first to report that Pritzker's wife and children, who had spent the last seven weeks at the family's $12 million horse farm in Florida, had moved to another family horse farm in Wisconsin, just across the Illinois border.

"Now I'm hearing that Gov. Pritzker has a 1,000-acre horse farm in Kenosha, WI," she tweeted. "That's where the family is tending to the animals tonight. You know, essential workers while the rest of us have been deemed 'non essential.'"

Pritzker later admitted his wife and children were not in Illinois, but claimed they were "essential" to work on his Wisconsin farm, which he said has livestock. Right. The billionaire's family was actively engaged in farming.

Three days after she broke the news, Pritzker revoked Jacobson's access to his daily press briefing. Pritzker's press secretary, Jordan Abudayyeh, criticized Jacobson for her attendance at a downtown Chicago rally of Pritzker critics.

"This weekend you attended and spoke at a political rally to fire up the crowd in opposing the Governor's policies to combat COVID-19," Abudayyeh said. "An impartial journalist would not have attended that rally in that capacity and therefore you will no longer be invited to participate as an impartial journalist."

Pritzker's suggestion that he only allows "impartial" journalists to attend press briefings singled Jacobson out, but not other Illinois reporters whose public opinions had been supportive of his policies.

Rich Miller of Capitol Fax has been the state's most prolific statehouse opinion writer and analyst for the past three decades. He attended Pritzker's daily press briefings— and remained a vocal supporter of the governor's statewide lockdown. In his weekly syndicated column, Miller applauded Pritzker's decisions and advised him on how to better communicate with Illinoisans and mitigate criticism.

Former president Barack Obama went golfing at the Robert Trent Jones golf club where he is a member. The club is 40 miles from his home, and he was driven there by a government chauffeur. There was no one around since all citizens of Virginia and Washington D.C. were under quarantine and are not able to play golf, so Obama and his friends had free reign of the course.

Two days later Washington D.C. residents received public service announcements on their cell phones that had been pre-recorded by Michelle Obama who was asked to do this by local government officials. Her message: "Remember we urge you stay home except if you need essential healthcare, essential food or supplies, or to go to your essential job."

Apparently, golf is essential – for the former president.[9]

Next Steps

Several strategies were announced for addressing the "constant threat" of the fake pandemic, but the two most important were testing and contact tracing, which were designed to control the spread of the disease, and "re-opening plans" which were supposed to provide a blueprint for resuming life in what we were told would be "the new normal." These were accompanied by an incomprehensible set of rules and regulations, with more rules added regularly to maintain control over the people. The governors had reason to be concerned. A few hundred million people anxious for some semblance of normalcy started venturing out, talking to one another, opening their businesses, patronizing favorite retailers, and returning to church.

Elected Officials Declare Themselves Rulers of People

What happened next was truly frightening. Governors of many states declared that they were above the law. They announced that they would veto any bills passed by the legislature that would reduce their authority. They refused to provide data when Freedom of Information requests were filed. And it was surprising how fast some of them made the transition from elected officials to rulers of subjects. One of the best examples took place in Ohio, where we live.

Mike DeWine is a former county prosecutor, state senator, member of the U.S. House of Representatives, lieutenant governor, U.S, Senator, and Ohio Attorney General. He practiced law for a while but has spent most of his career in politics. He was elected governor of Ohio in 2018 and declared himself the Emperor of Ohio in March 2020.

His behavior and decisions during this debacle were a major departure from a guy who was featured in a *Columbus Monthly* article titled "Mike DeWine: Making Ohio Nice Again." The article stated that lawmakers, political allies, and even those who were opposed to his positions referred to him as "gracious, friendly, pragmatic – even nice."[10] We voted for him, and for two years we thought we had made the right decision. Then along came COVID-19.

We gave then-governor DeWine the benefit of the doubt when he locked the state down in early March. He presented what was to be a short-term strategy that would allow health officials led by Dr. Amy Acton to assess the COVID-19 threat and make appropriate plans.

Immediately then-Governor DeWine attracted an enormous amount of national attention, appearing on news programs on almost every major network. CNN described him as "one of the most proactive governors" who exhibited a "...calm tone and steady focus on transparency" when he became one of the first governors in the country to shut down almost all activity.[11]

At the time this book is being published, DeWine still enjoys a favorable national profile and is lauded for the decisions he and his former sidekick Acton have made ostensibly to "protect Ohioans" and "save lives." But DeWine's decisions, starting with putting the lives of 11.6 million Ohioans under the management of a health director with a questionable background, poor judgment and limited capabilities, are ruining the lives of millions of Ohioans and have destroyed life as we knew it.

The Road to Hell Began in Mid-March

On March 9, 2020 DeWine and Acton declared that Ohio was "under a state of emergency." At that time three people had tested positive for COVID-19. Businesses were gradually closed over the next several weeks, until only those deemed essential could remain open. These were mostly grocery stores and big box stores like Walmart. Restaurants, bars, gyms, arts organizations, libraries, schools, and everything else was shut down. Millions of people were put out of work, and many businesses failed due to the length of the lockdown. Suicides increased, drug overdoses increased, violence increased, food insecurity increased, and homelessness increased. Reports indicated that most children were not "schooling at home" since a significant percentage of children did not even have an internet connection that would allow them to access a teacher or assignments. Almost all medical care was suspended. All of this was justified by Acton and DeWine as acceptable collateral damage to "flatten the curve."[12]

On March 17, 2020, DeWine and Acton cancelled Ohio's primary election without calling a special legislative session as required by law.[13] This was the date former Governor DeWine started to transform himself into the Emperor of Ohio.

According to some of our contacts in the legislature, the emperor made it clear that if the legislature passed any bills intended to reign in his power, he would simply veto them.

Was Any of This Justified?

Acton used the discredited Imperial College of London model to estimate that at the time the debacle started there were already 100,000 Ohioans infected with COVID-19. She stated that there would be 62,000 new cases of COVID-19 per day in Ohio, until as many as 70% of Ohioans would be infected. She issued an emergency order, which DeWine used to justify a two-week shelter-at-home order on March 23, 2020. At that time, DeWine promised Ohioans that he would use "...the best science, medicine and data, and to deploy all necessary resources to flatten the curve..." He also said that his administration would be transparent with all data.[14]

Ferguson's model was shown to be significantly wrong, off by 96%. Yet Acton revised her projections only twice, once to 10,000 cases per day and the second time to 2000 per day. With the exception of one prison at which all prisoners were tested, the number of cases never exceeded 100 per day.[15] In other words, Acton's continued projections remained 20 times higher than the actual infection rate, and these inflated numbers were used to continue the state's draconian measures, including a ban on visiting nursing home residents, religious services and gatherings, and even small groups in private homes.

During the daily press conferences, which we watched until we could not stand it anymore, data on cases and deaths were reported, or rather misreported. For example on April 14, 2020 it was reported that 50 people had died during the past 24 hours when there actually were only five, and on May 23, 2020 the Ohio Department of Health reported 84 deaths during the previous 24 hours when there were only seven.[16]

In addition to reporting false data on deaths, DeWine and Acton misled Ohioans by not disclosing who was dying. Acton insisted daily that nursing home deaths constituted only 20% of the total, but she must not have looked at any of the data before making those statements. By May 21, 2020 it was clear that

192

there was no reason to continue restricting the activities of most Ohioans since 79% of the state's deaths had been nursing home residents.[17] Nevertheless, the restrictions continued into the summer. This fall Ohio schools will not be reopening normally, arts organizations remain shuttered, sports events are prohibited, and masks are required when leaving the house.

A Troubled Person in Charge

In our opinion, one of DeWine's many questionable decisions was putting Acton in charge. In a televised press conference, Acton's demeanor appeared much like that of a celebrity author at a Barnes and Noble book signing when she locked down the state until April 30. 2020. During this and many other press conferences, she seemed totally disconnected from the massive destruction she continued to inflict on the lives of Ohioans.

According to an investigation conducted by Operation Rescue, Acton has suffered from mental health and addiction issues. On her medical license application she checked off "yes" to this question: "Have you ever been treated but not hospitalized for emotional or mental illness, drug addiction, or an alcohol problem?"[18] Operation Rescue tried to obtain further details from the Ohio Department of Health (ODH), but was denied while Acton was still director.[19] To be clear this alone should not have disqualified Acton from being appointed Director of ODH. But this, along with other information uncovered by the Operation Rescue investigation, should have been considered when giving Acton control over the lives of 11.6 million people for several months.

Acton has shared her personal background publicly, claiming that she was neglected as a child, experienced periods of homelessness, and often did not even have food to eat. She says that criminal charges were filed against her mother and her stepfather and indicated that she was abused. She survived all of this, she says and went on to become a medical doctor who was passionate about saving the world. In response to her story, DeWine once said, "She's special, isn't she?"[20]

But her mother, Donna Arthur, says that much of this is not true. Acton was never homeless, and always had food. The Operation Rescue reporter was able to validate Arthur's claims by checking her education and work history. There is no evidence that criminal charges were filed against Arthur or her husband in Mahoning County, where Acton lived at the time.[21]

To be clear, a background of addiction and mental illness should not disqualify anyone from public service or from holding most jobs. But the use of models proven to be wildly inaccurate, the refusal to adjust predictions, the reporting of false data; and instituting and maintaining a public health policy that caused harm to more people than could possibly be helped are signs of gross incompetence and someone who was not capable of doing the job for which she was appointed. Her mental health and addiction issues might explain her inappropriate demeanor and the appearance of being disconnected from the consequences of her actions. Certainly, DeWine had the resources to investigate all of this before appointing her. As her boss, he should have been monitoring her performance and fired her as it became clear that she had no idea what she was doing.

Acton resigned on June 17, 2020.

Why Are Lawsuits Required in Order to Obtain Data?

DeWine pledged to be transparent when he started taking away the rights of Ohioans. He was anything but. Reporters repeatedly asked the administration to provide information about the number of deaths in long-term health facilities. ODH refused, citing "privacy concerns" but offered no real legal justification. In the meantime, reporters started checking with nursing homes to confirm data posted on the ODH website. For example, ManorCare Health Services in Parma was reported to have 63 cases and the actual number was 36.[22] Deadline after deadline for posting accurate numbers was missed.

The American Policy Roundtable requested information from the DeWine administration about models and projections used by ODH and Ohio State University to project the spread of COVID-19 and make policy decisions. By law, these records must be disclosed in response to a Freedom of Information Request (FOIA). ODH denied the request, claiming that the department had immunity from all FOIA requests until the end of the pandemic.[23]

The Emperor has set up a perfect system for ruling his subjects. He publicly stated that he would veto any attempt by the legislature to reign in his power and blocked all access to data that could invalidate his claim that his policies are "saving lives."

It's Not Over, According to the Emperor

While the CDC reports that deaths have been falling for eleven weeks and that COVID-19 is no longer a pandemic,[24] the Emperor had decreed that masks be worn statewide. He has threatened to lock down the state again if cases continue to rise. He recently ordered all bars and restaurants to stop serving alcohol after 10PM, which will cause even further economic devastation in an industry that was barely hanging on before this edict.

The data supporting this? On Wednesday, July 15, 2020, Clermont County in Ohio was designated as "red," meaning that the county was classified as having a "public health emergency." One of our members called Clermont Mercy Hospital and found out that there were two COVID patients in a hospital with 147 beds and 16 ICU beds. This is considered overwhelming?

On Wednesday, July 15, 2020, there were 160 COVID patients in the hospital, statewide;[25] hardly a reason for concern in a state that has over 27,000 available beds.[26] New cases on Thursday July 16, 2020, including those classified as "probable" which means that they are not confirmed: 1093.[27] Many of these people were likely asymptomatic since the state was sending large numbers of people into neighborhoods to knock on doors and randomly test people.

As of summer 2020, it appears that the Emperor is intent on continuing to ruin the Ohio economy, destroy businesses, deprive children of education, increase homelessness, maintain a police state and issue edicts regularly until external intervention takes him out.

Unfortunately, DeWine Was Terrible, But Not the Worst

Governors in other states issued even more draconian rules than DeWine, with severe consequences for anyone who dared to disobey. The behavior of some governors, namely Kate Brown

(Oregon), Jay Inslee (Washington), Gavin Newsome (California), Andrew Cuomo (New York), Tom Wolf (Pennsylvania), Jay Pritzker (Illinois), Gretchen Whitmer (Michigan) and Tim Walz (Minnesota) was particularly egregious.

What was interesting about all of the governors who locked their citizens down, closed businesses and schools, mandated masks, shut down parks, sent police to patrol beaches, was that while they may have been unpopular with some citizens in their states before this debacle took place; they had never exhibited signs of being control freaks or being evil people. After declaring a public health emergency, which gave them unprecedented power, they became increasingly hostile and tone deaf to the suffering of their people.

Autumn Wehr, age seven, was admitted to Children's Hospitals in Columbus, Ohio after a prolonged seizure due to a brain tumor and was placed on a ventilator. DeWine had ordered that only one parent could be with a child during each 24-hour period due to COVID-19. The hospital refused to change its policy without permission from DeWine and DeWine refused to even respond to a direct request from the parents. What kind of person refuses to allow both parents to be with their dying child?[28]

When confronted with protesters who had not received unemployment checks and were running out of money, Governor Cuomo said, "You want to go to work? Go take a job as an essential worker. Do it tomorrow."[29]

Cuomo also threatened the Orthodox Jewish community with police action if they held large funerals, even for prominent rabbis.[30]

The behavior of these people, who ceased to act like elected officials and more like emperors and empresses, is best explained by Philip Zimbardo, in his book *The Lucifer Effect: Understanding How Good People Turn Evil.* Zimbardo is best known for conducting the Stanford Prison Experiment, which involved randomizing male college students to play the role of guards or prisoners in a mock prison. Within only a few days, the experiment had to be stopped because students, who had been tested and determined to be psychologically healthy before the experiment began, had become either brutal sadistic guards or emotionally devastated prisoners.

His book is about people who have done very bad things to other people, often claiming a high sense of purpose and moral imperative. He shows how a person's character can be changed through situational forces, particularly power, and how everyone is susceptible to crossing over to "the dark side."[31]

Zimbardo was an expert witness in the court martial hearings of the army reservists who were accused of criminal acts at the Abu Ghraib prison in Iraq. And he is likely to be called upon to testify if criminal cases are filed against some of the governors and health officials who made decisions that resulted in severe emotional trauma, physical illness, and death.

Lockdown Madness

Some of the rules and regulations issued by state health officials were almost hard to fathom. Here are just a few examples:

Alaska health officials mandated that religious services could be held outdoors, but only if the cars were parked six feet apart, and people remained in their cars at all times. Churches were ordered to have clearly marked parking stalls, or to have parking lot staff wearing reflective clothing and masks or other face coverings to properly direct traffic. If singing was included in the service, then cars were to be parked 10 feet apart.[32]

Alaska officials also issued strict instructions for Easter basket assembly:

- Anyone assisting with basket assembly or distribution must be screened and not allowed to participate if they meet any of the following criteria: a) have a fever, cough, shortness of breath, or other symptoms of respiratory infection; b) have a history of out-of-state travel within the past 14 days; or c) have a history of close contact to a person with COVID-19 or an undiagnosed respiratory infection in the past 14 days.
- No gathering may be of more than 10 people and a minimum of six feet must be between every person included in assembly and distribution of baskets.
- Wash hands with soap and water for at least 20 seconds prior to and after handling baskets or basket contents.
- Maintain at least six feet or more distance from people other than household members.
- Wear a cloth face covering when around people other than household members[33].

In July, Emperor Newsome of California decided that businesses that had been closed for months and then reopened for a short while could serve the public if they did so outdoors. Salons, barbershops, nail salons and even massages were ordered to provide services outside. The Emperor declared that to avoid over-exposure to sun, tents and canopies could be used as long as only one side was closed to allow for "sufficient outdoor air movement." Employees and clients would be required to wear masks, stations would need to be six feet apart and frequent disinfecting of all surfaces was mandated.[34]

Emperor Cuomo of New York State joined Emperor Newsome in ordering restaurants to offer outdoor dining. He also stipulated that people could not order only alcoholic beverages when patronizing restaurants and bars. Food had to be ordered, too, and the amount of food was prescribed – it had to be enough to constitute a meal. In California chicken wings, cheese sticks, fried calamari and French Fries were not considered meals. After being commanded by the Emperor to issue specific guidelines on what constituted a meal, the California Alcoholic Beverage Control Agency started posting these rules:

> It is often easier to describe what does not constitute a bona fide meal. In that regard, while the statute excludes mere offerings of sandwiches and salad, the Department does recognize that many sandwiches and salads are substantial and can constitute legitimate meals. Once again, the Department looks at the totality of circumstances and generally considers that

pre-packaged sandwiches and salads would not typically meet this standard. In addition, the Department will presume that the following, and offerings similar to them, do not meet the meal requirement:

- Snacks such as pretzels, nuts, popcorn, pickles, and chips
- Food ordinarily served as appetizers or first courses such as cheese sticks, fried calamari, chicken wings, pizza bites (as opposed to a pizza), egg rolls, pot stickers, flautas, cups of soup, and any small portion of a dish that may constitute a main course when it is not served in a full portion or when it is intended for sharing in small portions
- Side dishes such as bread, rolls, French fries, onion rings, small salads (green, potato, macaroni, fruit), rice, mashed potatoes, and small portions of vegetables
- Reheated refrigerated or frozen entrees
- Desserts[35]

Guidelines were even developed for sex. The British Columbia Centre for Disease Control created an entire section of its website called "COVID-19 and Sex," which featured this advice:[36]

You are your safest sex partner. Masturbating by yourself (solo sex) will not spread COVID-19. If you masturbate with a partner(s), physical distancing will lower your chance of getting COVID-19.

Virtual Sex:

Video dates, phone chats, sexting, online chat rooms and group cam rooms are ways to engage in sexual activity with no chance of spreading COVID-19. Be aware of the risks of sharing information or photos online, and web camming. Some people do not share personal information or show their face or other identifiable body parts, for more privacy.

Sex with partner(s):

Having 1, or a few, regular sex partner(s) can help lower the chances of being exposed to COVID-19. Talk with your sex partner(s) about:

- The types of sexual activities you want to have with them, and
- The precautions that you can each take to make sex safer for you and your sex partner(s), like wearing a mask and social distancing, and
- Whether you or your sex partner(s), or anyone you are in contact with, have a higher chance of getting a more serious COVID-19 illness (such as someone with an underlying medical condition like diabetes, lung disease, cancer or a weakened immune system)
- Before and after sex:
  - Wash your body with soap and water.
  - Wash your hands with soap and water for at least 20 seconds.

o Wash sex toys thoroughly per the manufacturer's instructions. Most, but not all, can be cleaned with mild unscented soap and water. Do not share them with multiple partners.

To Protect Yourself During Sex:

- Wear a face covering or mask. Heavy breathing during sex can create more droplets that may transmit COVID-19.
- Avoid or limit kissing and saliva exchange.
- Choose sexual positions that limit face-to-face contact.
- Use barriers, like walls (e.g., glory holes), that allow for sexual contact but prevent close face-to-face contact.

Some Businesses and Individuals Took the Regulations Seriously!

One of our viewers wrote, "My husband went to Valvoline near where we live in MA, to have the oil changed in our daughter's car. It's a drive through one where you wait in the car while they change the oil. He drove in and they asked him to wear a mask and he said he didn't have one but would roll up the windows. They said no they couldn't change the oil if he wasn't wearing a mask!! Where is the logic in that! A mask stops the spread but being enclosed in glass and steel doesn't do the job!!!"

Another reported this experience: "The other day I went into one of our small local shops to use the self-service machine to print off a label and send a parcel. The shop has only recently reopened and of course there are Plexiglass screens and an air of siege about the place (a little ironic since it sells alcohol products which if consumed in excess are probably far worse for your health than any virus!).

"Anyway there were only two people in the place - me as a customer and the assistant, who by the look on her face was clearly taking everything very seriously and was enjoying her newfound sense of power in directing the public.

"I walked over to the machine, and as I started to type in the information, she said "For future reference, there are direction arrows on the floor, could you walk round following them?" Indeed, on inspection, there were, which I had not noticed. "But there are only two of us in the shop" I said, "it's irrelevant." "No" she replied, "it's very important!" "No it isn't" I replied, 'there's no one else here!"

"She then said, in a very official tone, 'Well I'll let you off this time but can you remember for next time?" Needless to say, I may go in to use the machine, but I won't be spending my money there in the future! What sanctions she proposes to apply I'm not sure, but a nasty glare may be in order!"

This is a classic example of mindless adherence to rules which make little sense but make it look as if you are doing something important. It makes me think that the only people to come out of this charade smiling will be the manufacturers of Plexiglass!

Another person wrote, "My son went out to a store for the first time in weeks and he went without a mask. The cashier said she couldn't serve him if he didn't have a mask on. He was incredulous. He asked if they sold them and she replied they were out of masks. He put the items down and left thinking he spent more time talking to her about not having a mask than if she just rang him up. He grew so irritated on the way home he turned around and went back, took his sweatshirt off, tied it around his face and went back to the cashier and asked, "Is this okay now?" She actually said no! A manager or coworker nearby overruled her and said it was ok. Ridiculous!"

Finally, a man reported that when he visited a drive-in movie, he was told that he would need to park six feet away from other cars, stay in his car (there were no concessions and restrooms were closed) and wear a mask while watching the movie.

The Resistance

The Apple Bistro in Placerville, California declared that its employees would not wear masks and that service would be denied to any customer who insisted on wearing a mask. A sign on the door stated: "Attn Government Agents. Please provide lawful and necessary consideration to aid the bearer in the unimpeded exercise of constitutionally protected rights. Thank you for your understanding and assistance."[37]

Big Walnut Local Schools in Central Ohio decided to have a normal graduation for 250 students even though it violated the rules and the school leaders could end up in jail.

"I find it insanely illogical that if we brought two cows and three goats to graduation and called it The Big Walnut Community

Education Fair Celebration, that it would be a completely legal event," commented board president Brad Schneider.

While others said they needed to set an example for students and called pursuing the ceremony an unnecessary risk, board member Doug Crowl made the motion to go through with it.

"If you want to name someone that they can come arrest, I'm willing," Crowl said. "What lesson do we want to teach (students)? Do we want to teach them that they are Americans, and they are adults?"

In a four to one vote, the board decided to break state orders and schedule the ceremony.[38]

An entire county in Colorado declared that it would open without permission when Weld County attorney Bruce Barker declared that business closures were not lawful. He said that unelected health officials had no legal basis for issuing orders and cited a Wisconsin Supreme court ruling that dismissed similar orders by that state's governor as justification for his stance. Barker also noted that the Colorado State Attorney General's office had not prosecuted a single case, most likely because the orders were not enforceable.[39]

Law enforcement also started pushing back. Over 60 sheriffs in over a dozen states publicly stated that they would not enforce lockdown measures.

For example, **Sheriff Chad Bianco of Riverside County California** told county commissioners on May 5, 2020, "I refuse to make criminals out of business owners, single moms, and otherwise healthy individuals who are exercising their constitutional rights."

Sheriff Scott Jenkins of Culpeper County Virginia decided not to enforce Emperor Northam's directives. He says that in the beginning, he supported Northam's orders since there were so many unknowns but as the lockdowns dragged on for months decided that they violated people's rights. "The governor does have the right to enact regulations during a state of emergency and I get that, but the Constitution doesn't go away just because of that," Jenkins said. "We won't be used to enforce an edict or regulation by a governor, health director, or anyone else."

"What is vital to a person's life or well-being?" Jenkins asked, discussing whether the disease is a great enough risk that we are willing to trample on the First Amendment our country was founded on. "They could easily argue that there's a bigger risk to get hit by a car."

Jenkins spoke passionately about religious freedom, stating, "It is not our place to step in and intervene in their worship."

Sixty-five thousand residents of Chaves County in New Mexico were essentially given permission to violate the Empress Grisham's orders without fear of punishment. "My department will not be out citing anyone for not wearing a mask," **Mike Herrington, the county sheriff**, told The Marshall Project. "I will not be enforcing any of those orders." Herrington had already allowed businesses to reopen in recent weeks.

The county includes Roswell, the town famous for a UFO sighting in the 1940s, and usually the host of a UFO festival that brings in a million dollars to local businesses. Of course, it was cancelled in 2020.

Herrington said he was approached by small business owners, some of them in tears. They complained that Walmart and Target, allowed to stay open because they sold food, were also selling televisions and other big items to residents looking to spend their stimulus checks. These owners told Herrington that by opening they could reduce crowding at the big box stores.

The head of the New Mexico Sheriffs' Association wrote a letter to Attorney General William Barr, asking him to consider whether the governor's orders violated the civil rights of New Mexicans, stating that the forced closure of businesses benefitted the big box stores. "These are my friends and family," Herrington said. "To look at the fear in their eyes, the fear of losing everything they have, tells me I have no choice but to stand and take on this fight."

Sheriff Bob Songer of Klickitat County California says that the job of the sheriff is not just to put bad guys in jail, but to "protect the liberties and God-given constitutional rights of the citizens they work for."

Former Arizona sheriff and founder of CSPOA Richard Mack told *The Epoch Times* that under the current circumstances, he believes that citizens haven't broken the law and that "they have only tried to maintain their business and their liberties as guaranteed in the Bill of Rights." He said the lockdown orders have not passed "through any state legislature."

"For any governor to appoint themselves as dictator over the state, to destroy people's businesses, to destroy churches, and to arrest people going to church or arrest ministers is an absolute outrage," he said. "I'm absolutely petrified as to what's going on here, all in the name of taking care of people." Mack pointed out that the number of deaths attributed to smoking exceeded the number of deaths from COVID-19 in the U.S. "We don't even make smoking against the law," he said. "I don't understand the inconsistency."

One of the reasons so much resistance came from sheriffs is that they are elected by the local population and answerable to that population. In the case of Herrington, Empress Grisham issued a press release asking New Mexicans to follow her orders. There is nothing she can do to punish Herrington.[40][41]

Butler County Ohio Sheriff Rick Jones held a press conference to announce that he would not be enforcing mask laws, stating, "I want to make sure everybody understands I am not the mask police. I am not going to enforce any mask wearing. Not my responsibility, not my job, people should be able to make this decision themselves."

In response to the governor's allegations that the hospitals were full of COVID-19 patients, Jones called the hospitals himself and was told that this was not true. He went on to say that if the health department chose to enforce the laws, officials could "...put a little yellow light on their cars and stop people." He went on to say, "They will not like the response. People are tired of this and worn out, and then added, "I've sworn to support the U.S. constitution and Ohio Constitution – not to the governor."[42]

Some rational voices in government also decided to fight back. **Governor Brian Kemp** banned cities and counties in Georgia from requiring citizens to require masks in all public places.[43] Unfortunately he was the exception, not the rule.

And then there were lawsuits filed – thousands of them!

They were filed by churches, businesses, groups of businesses, elected officials, and citizens. The outcomes were variable, but in many instances the courts sided with the plaintiffs. The usual response from the emperors and empresses was to appeal the decisions to maintain their power. Thousands of cases are still pending, and more are being filed each day.

Dire Consequences for Those Who Disobeyed!

The ginned up frenzy resulted in destruction of property and violence in some areas.

First Pentecostal Church in Holly Springs, Mississippi filed a lawsuit against the city over its public health order. City Attorney Shirley Byers says that the church was issued a violation on April 10, 2020 after 40 people had gathered inside and were not social distancing. The police disrupted Bible study and Easter Sunday Service. The lawsuit says that social distancing is practiced, and that the congregation is only moved inside when weather prohibits being outdoors.

On May 22, 2020 the church was burned to the ground and "Bet you stay home now you hypokrits" was painted in the parking lot of the church. It is being investigated as an arson. Interesting that whoever did this is calling the church members hypocrites when the arsonists had to leave home to burn the church down.[44]

Two owners of a New Jersey gym were arrested after defying Governor Phil Murphy's stay-at-home and busines closure

210

orders. The partners were issued numerous citations and at one point the health department boarded up the doors of their business. All they were doing was trying to make a living. Ian Smith and Frank Trumbetti kept detailed records of every person who visited the gym, and thousands of people did. Not one person was reported to develop COVID-19. The emperor of New Jersey requested that the court allow law enforcement to take "extraordinary measures" to stop them from continuing to operate their business.

"I'm not afraid of tyrants. No American should be," Smith said. "We outnumber them greatly and the only thing that they run off of is fear, which is why you see what you see in the media where they are pumping fear ... They don't ever offer any solutions. It is, 'Wear a mask, shut up, and wait for a vaccine.'

"That's not public health," Smith asserted, "and I won't subscribe to it."[45]

Jose Freire Interian and his wife tested positive and were ordered to quarantine themselves and wear masks at home for 14 days. The Florida Department of Health determined that they presented "an immediate danger of harm to others" when the agency found out that the couple had the nerve to go shopping for food. A neighbor captured footage of the couple walking their dog while still under the quarantine order and sent it to police, who asked for and got a warrant for their arrest.[46]

Elizabeth Linscott, who lives in Hardin County Kentucky, decided to get tested for COVID-19 before driving to Michigan to visit her parents and grandparents. She tested positive but had and continued to have no symptoms. The health department contacted her and told she needed to sign a Self-Isolation and Controlled Movement Agreed Order, forcing her to agree not to go anywhere without calling the health department first. She refused.

Two days later eight people showed up at her house in five cars including a man from the health department wearing a suit and mask, and delivering three papers to her, her husband, and her daughter. Elizabeth and her husband were ordered to wear ankle monitors and if they moved more than 200 feet, law enforcement would be notified.

The health department petitioned a judge to get the order. The Linscotts were planning to hire an attorney to fight the action in court.[47]

A man in Spokane county who refused to isolate after testing positive for COVID-19 was jailed. The man was issued a Civil Health Hold by County Health Officer Dr. Bob Lutz. "His unwillingness to cooperate," Lutz says, "means that he is a direct threat to the community."

Apparently, the man refused to take the "more appropriate housing" which the county offered to him, which included hospitalization with guards or a "more appropriate isolation facility." They were left with no other choice but to take him to jail, where he was held under the medical care of the jail.

He was sentenced to stay until July 11, 2020, unless Lutz decided there was "need for a longer quarantine." If this was the case, he planned to petition the court and send the order to the jail.[48]

The Emperor of Arkansas, Asa Hutchinson, mobilized the National Guard to transport COVID-19 patients from their homes to an isolation facility near the University of Arkansas in Little Rock. This was to allow people who live with others to be properly isolated if they tested positive with one of the fake tests. Members of the Guard were ordered to work 12-hour shifts in order to make sure that coverage was available 24 hours per day, which means that some abductions – sorry we meant to say "transports" – would be done in the middle of the night.[49]

On July 17, 2020. Ventura County California forcibly locked residents inside a seven-story public housing complex after a resident was hospitalized and tested positive for COVID-19. Common areas were shut down, keys were disabled, and residents were not able to get back in if they decided to leave and tried to return home. Even the laundry area was closed. All residents were placed under house arrest until all could be tested and results returned. Testing was mandatory and no one was permitted to opt out. After several days, one more resident tested positive, and was quarantined and the rest were released from house arrest.[50]

The Most Hated Rule: Masks

Mask mandates were a lot like lockdowns. Once a few mayors and commissioners and governors started mandating them, the domino effect began. Every few hours a new locale mandated that citizens wear masks. The mandates became more and more strict, with threats of fines or closure for businesses that allowed anyone to enter without a mask, and citations and even jail for citizens who chose not to wear a mask whenever they left their homes.

Large retailers started mandating masks nationally in all locations, even those located in the few remaining places that did not require masks. These included Starbucks, Target, CVS, and Walmart. Walmart decided to station "health ambassadors" dressed in black at store entrances to "educate" customers about the importance of keeping everyone safe by wearing a mask.

The frightened people who believed that there was truly an epidemic became militant about masks. Local governments set up snitch lines to report anyone not wearing a mask and the mask wearers took their jobs seriously! They took pictures and made reports. Fights broke out in stores.

A security guard working at a Target store in Van Nuys California ended up with a broken arm after a confrontation with two men who refused to wear a mask when entering the store. According to local law enforcement this was the fourth fight over masks reported at this store since the mandate was made.[51]

A woman was arrested after a disagreement about masks between two customers turned into a physical fight in Illinois.[52]

A man was shot by a sheriff's deputy after stabbing another man who screamed at him for not wearing a mask at a convenience store in Michigan.[53]

Many citizens were more outraged about being forced to wear masks than they were about locking down businesses.

In view of the requirement for almost all citizens to wear masks almost all of the time, it is reasonable to ask whether or not masks are effective for preventing the spread of COVID-19, as well as the potential health effects of wearing a mask for several hours every day.

OSHA's Policies Regarding Masks

The Occupational Safety and Health Administration (OSHA) has published policies concerning the wearing of masks in the workplace and took the extra step of establishing guidelines for mask wearing for COVID-19. OSHA offices are in every state and the rules concerning masks vary from state to state. An OSHA Bulletin pertaining to "respiratory protection guidance for employees and workers" posted on the website for the United States Department of Labor includes these guidelines (italicized emphasis ours):

Are there any cautions or limitations when using respirators?

Yes. Each type of respirator can come in several varieties, each with its own set of cautions, limitations, and restrictions of use. *Tight fitting respirators require fit testing to ensure an adequate fit to the face, and cannot be used with facial hair.* Certain escape respirators use a nose clip and mouthpiece, which is clenched between your teeth, similar to a snorkel. Some respirators prevent the user from talking while others have speaking diaphragms or electronic communication devices. Every respirator contaminated with hazardous chemicals should be cleaned and decontaminated or disposed of properly.

All respirators require training in order to be properly used. Sometimes you can practice using your own respirator. Some escape respirators come in a package that must remain sealed until use, so you need to be trained using a special "practice" version.

Training is extremely important in regard to the storage, maintenance, use, and disposal of the respirator. This information is provided by the supplier of the respirator (i.e., seller, distributor, or manufacturer). If you do not use a respirator correctly, it is very likely that it will not adequately protect you and may even hurt you.

How well does a respirator need to fit me?

If your mask does not make a tight seal all the way around your face when you inhale, you may breathe contaminated air that leaks around the edges of the face seal. Most respirators come in different styles and sizes, and fit different people differently because people's faces have different shapes. You also need training to know how to correctly put the mask on and wear it correctly. This information should be provided by the supplier of the respirator.

The only way to tell if a tight-fitting respirator fits you properly, and is capable of protecting you, is to fit test the respirator. Fit testing can be accomplished a number of different ways and should be done by a health and safety professional before workers wear a respirator in a hazardous environment. Respirators must be checked for proper fit each time they are donned to ensure they provide adequate protection.

Can I wear a respirator if I have a beard?

Anything that prevents the face mask from fitting tightly against your face, such as a beard or long sideburns, may cause leakage. If your respirator requires a tight fit, you must trim back your beard so that it will not interfere with the face-facepiece seal. If your respirator is a loose-fitting (hooded) positive pressure respirator (e.g., a powered air-purifying respirator, PAPR) then you may have a beard.

If I have the right cartridges/filters for a certain hazard, and my mask fits, will they always protect me against that hazard?

No. Gas masks and respirators reduce exposure to the hazard, but if the exposure is such that it goes beyond what the filter is capable of handling (either because the amount of toxic gas or particles is more than what the filter is designed to handle, or because the exposure lasts longer than what the filter is designed to handle), the filter may not be effective in providing required protection. Also, there may be a small amount of leakage even if the fit of the respirator has been tested. If so, and if there is a large amount of a toxic chemical in the outside air, even that small leakage can be dangerous.

Can anyone wear a respirator?

No. Breathing through a respirator is more difficult than breathing in open air. People with lung diseases, such as asthma or emphysema, elderly people, and others may have trouble breathing. People with claustrophobia may not be able to wear a full

217

facepiece or hooded respirator. People with vision problems may have trouble seeing while wearing a mask or hood (there are special masks for people who need glasses). Employees must be medically evaluated before assigned to use a respirator.

Will my cartridge/filter and respirator mask protect forever?

No. Cartridges, filters, and masks get old. If the filter cartridges are outdated, have been open to the air or are damaged, you may not be protected. Cartridges that contain charcoal or other chemicals for filtering the air should be kept in air-tight packages until use. If cartridges are open or not packed in air-tight packaging, they should not be used. Even cartridges in original packaging have expiration dates that should be checked before purchase and use. Also, over time your mask can get old and break down. Keep your mask in a clean, dry place, away from extreme heat or cold. Inspect it before and after use according to the manufacturer's instructions. Cartridges also have a limited service life; they must be changed periodically during use.

Will a gas mask protect me if there is not enough oxygen in the air?

No. Air-purifying respirators do not provide oxygen. If used in an environment with low oxygen levels, such as in a fire or a confined space, you are in danger of asphyxiation.[54]

California's OSHA rules are even more specific, requiring medical clearance by a licensed healthcare provider before an employee can be permitted to wear a mask at work. Masks must also be fit-tested by trained personnel.[55]

OSHA's statements are clear: It is important to select a mask that has been proven to protect against the hazards encountered. It is not safe for all people to wear a mask. Some states require medical clearance. Masks must be fitted properly, replaced frequently, and disposed of properly. Anyone with facial hair cannot be protected with masks.

How to Wear a Mask

Most people who are wearing masks are not putting them on, maintaining them while on, nor disposing of them properly. Applying the mask should begin with washing the hands with soap and water for 20 seconds or using a hand sanitizer with at least 60% alcohol. The mask is then placed over the nose and mouth. The seal must be tight with no gaps between the face and mask. People should not touch their masks and if they do should wash their hands or use sanitizer immediately. It should be removed carefully without touching the front and discarded into a closed bin, and then hands should be washed again.[56]

Masks can be used again if they are taken care of properly. The first rule is not to touch the mask, according to Dr. Lucien Davis, an epidemiologist at the Yale School of Public Health. He says touching the mask can transfer virus particles to the surface. He also advises that masks should be removed in a safe place away from other people. Only the ear straps should be touched.

Jade Flinn, a nurse educator for the Biocontainment Unit at Johns Hopkins Medicine, says that used masks should be placed in paper bags with good ventilation for the mask to air out.

Both Flinn and Davis say that people who wear masks should have 3-5 of them so that they can wear one while the others are airing out.[57]

The best option is to throw away disposable masks in a plastic bag, avoid touching the edges of the bag, and then let it sit for a few days. Disinfectant is not recommended because it stays in the fabric and some are dangerous to breathe in.

It is almost a certainty that most people are not handling, wearing, or disposing of masks in a safe manner.

Are Masks Effective?

The COVID-19 virus is **0.125 μm** (micron) in size and can penetrate the surgical mask barrier. According the US National Academy of Sciences, in community settings "face masks are not designed or certified to protect the wearer from exposure to respiratory hazards."[58] Another study showed that surgical masks do not provide protection for aerosols ranging from 0.9-3.1 μm.[59]

In household settings, surgical masks do not prevent transmission of flu.[60] [61]

A study of eight different brands of face masks concluded that the masks did not filter 20-100% of particles between 0.1 and 4.0 microns.[62]

A review of 17 studies concluded, "None of the studies we reviewed established a conclusive relationship between mask/respirator use and protection against influenza infection.[63]

An article in the *Journal of the American Medical Association* in April 2020 advised, "Facial masks should be used only by individuals who have symptoms of respiratory infections such as coughing, sneezing or, in some cases, fever…Face masks should not be worn by healthy individuals to protect themselves from acquiring respiratory infection because there is no evidence to suggest that face masks worn by healthy individuals are effective in preventing people from becoming ill."[64]

The Food and Drug Administration classifies face masks as Class II devices, but when used for non-medical purposes are not regulated by the FDA. The FDA's website concerning masks used to protect against COVID-19 states that the FDA does not expect manufacturers of face masks "…to comply with certain regulatory requirement, where the face mask does not create an undue risk in light of the public health emergency." The site goes on to state that "..the FDA believes face masks…do not create such an undue risk where…the labeling does not include uses for antimicrobial or antiviral protection, infection prevention or reduction, or related uses, and does not include particular filtration claims." In other words, the FDA says masks are fine as long as they are not marketed with claims about protecting against viral transmission.[65]

Due to shortages, some people have been advised to make their own masks out of cloth. According to a hospital study in which hospital wards were randomized to medical masks, cloth masks, or a control group which included a high proportion of people who wore some type of mask, the rate of infection was highest in the cloth mask group as compared to the group wearing some type of medical mask. Transmission of viral particles through cloth masks was almost 97%, compared to medical masks at 44%.

Reasons cited included moisture retention, reuse of the masks, and poor filtration, all of which can increase rather than decrease the risk of infection. The researchers concluded that the results "... could be interpreted as harm caused by cloth masks."[66]

According to the CDC Masks Do Not Work

Health authorities and government officials claim they are "following the science" and often cite the CDC as a source of guidance. During the first few months of the COVID-19 debacle, the CDC and the WHO stated the masks were not effective for preventing transmission of viruses. A study posted on the CDC's website dated May 2020 reported the results of a review of ten randomized controlled trials that were published between 1946 and 2018 and that looked at the efficacy of face masks for reducing lab-confirmed influenza. The pooled analysis showed no benefit.[67]

What Does Fauci Say About Masks?

In March, Fauci told Americans during a 60 Minutes interview that they did not need to wear masks.[68] A few months later he changed his mind and said that the reason why he told Americans not to wear a mask was not because it was not a good idea – it was, he said, but it was because he needed to make sure that there were enough masks for health professionals. Thus, he claimed to have misled Americans for a good reason.[69]

During a conversation with students at the Georgetown University Institute of Politics and Public Service, Fauci was asked about conducting a study concerning the efficacy of masks.

"What kind of studies can we do right now in the middle of the pandemic about masks and transmission of the disease?" the student asked. "Or are we just relying on anecdotal evidence because we are not able to do those kind of studies right now?"

Fauci balked at the idea. "I would not want to do a randomized controlled study because that would mean having people not wear masks and see if they do better," he said. In other words, Fauci is comfortable telling everyone to wear a mask, but not conducting a study to determine if wearing a mask works.[70]

On July 29, 2020 Fauci announced that in addition to masks, people should start to wear goggles. According to Fauci the addition of goggles would result in all mucosal surface being shielded – the nose, the mouth, and the eyes.[71]

What is next? Bubble wrap?

Can Masks be Harmful?

"Maskme" is the name of the condition that develops because of wearing useless masks continually. According to Dr. Amer Jaber of Washington Square Dermatology in New York, "When you wear a mask, you seal in your breath. This creates a moist, humid environment as it traps your breath, skin oils and sweat, leading to irritation, rashes and acne." He compares the condition to diaper rash and says that the masks themselves may irritate skin just by being in contact with the face for long periods of time. This can further irritate eczema, psoriasis, or rosacea. He says, "The longer you wear the mask, the worse it is."

Jaber recommends taking off the mask whenever possible to let the skin breathe, and using lotions, creams and even Vaseline if necessary, to soothe irritated skin. Masks need to be aired out and dried after use and washed to remove oils and sweat trapped inside.

As for how to treat acne if it has already developed, Jaber recommends benzoyl peroxide, salicylic acid, or glycolic acid. He warns people, however, to be careful since topical acne creams can cause further irritation and worsen skin conditions resulting from the constant wearing of masks.[72]

In addition to acne and skin disorders, wearing masks can cause headaches, particularly in those who are predisposed. One study concluded, "Shorter duration of face-mask wear may reduce the frequency and severity of these headaches."[73]

A study of 158 healthcare workers age 21 to 35 showed that while only one third of the subjects reported pre-existing headache diagnosis, 81% reported PPE-associated headaches because of wearing masks four or more hours per day.[74]

Unfortunately, there have been much more serious consequences from wearing masks than skin irritation and headaches. A woman in New Jersey crashed her car into a power pole because she was wearing an N95 mask for several hours, and insufficient oxygen and excessive carbon dioxide intake caused her to pass out, according to the Lincoln Park Police Department. She was not under the influence of drugs or alcohol.[75]

Research shows that medical masks lower blood oxygen levels,[76] and increase carbon dioxide levels[77] and that even healthy people can develop symptoms of hypoxia, or low tissue oxygen. As oxygen decreases and carbon dioxide increases, the lungs try to compensate by increasing the breathing rate. This is generally an ineffective strategy, however, and some people develop symptoms of carbon dioxide toxicity.

It is dangerous for people who have Chronic Obstructive Pulmonary Disorder (COPD) to wear masks because they increase breathlessness.[78]

Wearing masks while exercising is even more dangerous. Two students died of cardiac arrest during physical education classes in China while wearing a mask.[79] Also in China, a jogger's lung collapsed as a result of running for 2.5 miles while wearing a mask.[80]

According to Dr. Jenny Harries, deputy chief medical officer in England, wearing masks can actually *increase* the risk of contracting the virus because the virus can be trapped in the material and cause infection when the person inhales. According to Dr. Harries, members of the public should not wear masks unless they are sick, and only if advised to do so by a healthcare provider.

"What tends to happen is people will have one mask. They won't wear it all the time, they will take it off when they get home, they will put it down on a surface they haven't cleaned," she said. Furthermore, people go out and about and don't wash their hands every time they touch something – they can't – and then touch their mask constantly to drink water or eat, or even communicate, and this becomes a means of infection.[81]

There is no clear proof that wearing masks is protective, and a considerable amount of evidence that doing so can be harmful. Research has never been conducted to examine the safety or long-term health consequences of wearing masks for most of a person's waking hours for weeks or months at a time. Essentially the government is conducting a clinical trial involving millions of people without their permission, without having subjected the protocol to a review board, without appropriate monitoring, and without any safety measures in place.

The better policy is to encourage sick people to stay home and allow healthy people to live their lives normally.

The rationale for all the mask wearing? More cases. COVID-19 was spreading through the population rapidly and had to be stopped and the only way to stop it would be widespread testing and contact tracing.

Why the Sudden Increase in Cases?

In an earlier chapter you learned that the tests used were approved without being properly scrutinized, and that the tests were highly inaccurate. But there's more.

When the pandemic first began, most of the people being tested for COVID-19 had symptoms. Due to several factors, tests were in short supply and most of the country was "sheltering in place." Once the curve seemed to have flattened and people were starting to leave the house again, tests became widely available. The general population, including those who were asymptomatic, was encouraged to get tested.

The price of the test varied depending on the type of test performed, the lab to which it was sent, and the location to which it was sent. According to one analysis of 78 hospitals, the range of PCR testing was between $20 and $850 per test. With a median of $127. A little over half of the tests cost $100-$199, and almost 20% were above $200. In addition to the test, insurance companies, Medicare and Medicaid were billed for an office visit and for collection of the specimen.[82]

Hospitals started offering testing, which was one of the ways in which they could recoup the money lost during the lockdown when almost all procedures were deemed "non-essential." Drug stores, urgent care and other outlets got involved in testing too because it was profitable.

States mobilized hundreds of people to knock on doors of residences to offer testing for free. We obtained a copy of a letter sent on letterhead from the Ohio Department of Health and The Ohio State University that included these statements:

"Your household was chosen as part of a larger group of households that may be visited by the project team."

"There is no cost to you."

"A team will be coming to your community during the week of July 9-12, 2020. If you do not want to participate, please contact us and we will remove your household from the list."

"The project will be recruiting participants seven days per week between 8 AM and 8 PM."

As you might imagine this is a great way to increase "cases," particularly with flawed tests. It is highly likely that most of the people tested were asymptomatic. Depending on the test used, a person who is asymptomatic could have tested positive due to having become infected several weeks prior, clearing the infection without developing symptoms (this happens every flu season to tens of millions of people) and still have harmless remnants of viral proteins present.

Reports to the CDC evidenced the significant increase in testing:

- Week of April 12-April 18 575,490 tests performed[83] 82,212 per day
- Week of July 19-July 25 1.906,631 tests performed[84] 272,375 per day

Not surprisingly, more testing yielded more cases.

The Council of State and Territorial Epidemiologists (CSTE) provided instructions for a new category of "case" – the "probable case" which was applied to people who had any symptom from a long list such as headache and a sore throat; who belonged to a "risk cohort"; or who had contact with anyone who tested positive such as living in or visiting "an area with sustained, ongoing community transmission."[85] While this vague information might be useful to an epidemiologist, it had absolutely no relevance to what would really be important to know, such as how many people were sick, the population most affected, and the mortality rate.

Armies of Contact Tracers were hired to follow up with "cases" to grow the case rate by asking people to disclose who they had been in contact with during the previous few days. These contact tracers were not required to have any medical training and were not even required to ever meet or even talk to people who were declared "cases." In other words, a person could be considered a case just by virtue of having been in contact with another person who had been declared a case, even if the original case was declared a case by virtue of contact with a person living in a place with lots of cases. Did you get that?

The cases could be categorized as a "probable" case even if the person did not answer the phone when the contact tracer called.

The CDC endorsed and promoted this byzantine scheme. Here are excerpts from the guidance document for contact tracers:[86]

"COVID-19 case investigations will likely be triggered by one of three events:

1. A positive SARS-CoV-2 laboratory test or

2. A provider report of a confirmed or probable COVID-19 diagnosis or

3. **Identification of a contact as having COVID-19 through contact tracing**

If testing is not available [or declined], symptomatic close contacts should be advised to self-isolate and be managed as a probable case. Self-isolation is recommended for people with probable or confirmed COVID-19 who have mild illness and are able to recover at home."

Not only did this inflate the number of cases in counties and states, it made accurate compilation of the data by the CDC almost impossible. For example, Arizona, Ohio, Michigan, and Virginia included probable cases and deaths in their reports. Some states, including Arkansas, New Jersey, and Washington included probable deaths but not infections.

Some states, like Maine and Kansas, included probable deaths but not probable infections. Other states reported probable cases or deaths versus confirmed cases or deaths or both separately, but the CDC listed both together in totals for those states. These included Alabama, Illinois, Massachusetts, Minnesota, and South Carolina.

Eight states decided to exclude probable cases and deaths from their totals, and those included Alaska, Georgia, Missouri, North Carolina, Nevada, and Oklahoma.[87]

There was a backlog of patients who had procedures and treatment postponed when hospitals were closed while waiting for the "surge" that never occurred. These patients were another source of cases. Almost all these people were required to have a COVID-19 test when admitted to the hospital regardless of the reason for hospitalization. This included people admitted for joint replacement, chronic conditions, injuries, and accidents. All these people who tested positive, regardless of whether or not they had symptoms, were considered "hospitalized COVID patients."

To further complicate matters, many of the hospital admissions coded as COVID-19 are probable rather than confirmed cases. For example, the Massachusetts Department of Health data showed that 70-80% of all COVID-19 hospital admissions were deemed "suspected" cases during the months of June and

July.[88] Thus, it is impossible to determine a true count of "real" hospitalized COVID-19 patients.

The federal government offered a tremendous incentive to code as many patients as possible "COVID-19." On July 17, 2020, the U.S. Department of Health and Human Services started distributing $10 billion to hospitals in "high-impact areas." This money was part of the CARES Act. Hospitals with over 161 COVID-19 admissions between January 1 and June 10, or one admission per day, or a "disproportionate intensity of COVID admissions" would be paid $50,000 per admission.[89]

Even worse, Robert Redfield, Director of the Centers for Disease Control, admitted that hospitals had a "perverse" incentive to increase their count of COVID-19 fatalities when he testified in front of a House panel on July 31, 2020. The reason being even higher reimbursement rates for deaths.[90]

Soon It Was Clear: Something Was Really Wrong with the Tests

Testing escalated and it seemed like states were competing with one another to see which could claim to be testing the most people. Soon it became apparent that these tests were almost worthless.

The Connecticut State Department of Health reported on Monday, July 20, 2020 that a flaw in COVID-19 testing resulted in at least 90 false positive tests. Between June 15 and July 17, 2020, 144 people were told they had positive results after their specimens were analyzed by Thermo Fisher Scientific in Waltham Mass.

Despite the 62.5% error rate, acting Commissioner of Health Dr. Deidre Gifford said that "Anybody who's received a positive test, they should absolutely assume that that positive result is correct until such time as they are informed by their provider of any change."[91]

In Texas state health officials reported a high incidence of false positive COVID-19 tests for both residents and employees at nursing homes in several locations.

For example, at the Snyder Oaks Care Center, 39 residents and employees tested positive. Most had symptoms so they were transported to a hospital and tested again and 100% of them tested negative. The same group that tested the residents and staff at this nursing home also collected specimens from first responders, law enforcement and people working at the local jail.[92]

In July, the Florida Department of Health reported that 22 labs reported 100% of all tests were positive, and two reported that 91.18% of tests were positive. This is statistically almost impossible. Additionally, 88% of tests at NCF Diagnostics in Alachua were positive. The positive rate was 98% at Orlando Health. When a television station contacted the center, the reporter was told that the positive rate was only 9.4%.[93]

As if all of this was not bad enough, on June 16, 2020 the FDA issued an update encouraging the use of pooled testing. This technique allows a lab to mix several samples together in a "batch" or pooled sample and then test the pooled sample.

This guideline allowed several samples to be tested together for the same cost as a single test. If the pooled sample is negative, the lab could conclude that all the subjects were negative. If the pooled sample tests positive, then each sample is tested individually to find out which was positive.

The benefit is lower cost because fewer tests are run overall, fewer testing supplies are used, and in most cases, the results can be returned to patients more quickly. However, because samples are diluted, there is less viral genetic material available to detect, which can lead to a greater likelihood of false negative results, particularly if not properly validated.[94]

Last but not least, a growing number of people started reporting that they scheduled a COVID-19 test at a testing location, and either never showed up, or that the line was too long and they left without testing. Yet these people received a notice from the center that they had tested positive. At the time that this book was being completed, we had confirmed and verified two cases in which this had occurred.

The bottom line: Out-of-control governors who had declared themselves free of any meaningful oversight found it easy to justify any action they chose by extending the "State of Emergency" based on made-up cases, many of which resulted from flawed tests.

ENDNOTES

1. Adolf Hitler. *Mein Kampf* 1926 https://history.hanover.edu/courses/excerpts/111hitler.html accessed 5.18.2020

2. The Power of Propaganda https://www.facinghistory.org/resource-library/teaching-holocaust-and-human-behavior/power-propaganda accessed 5.18.2020

3. Laura F. Deutsch. Purim at Goebbels' Castle. *Aish.com* Mar 4 2020 https://www.aish.com/h/purim/t/ts/Purim-at-Goebbels-Castle.html accessed 5.18.2020

4. http://libertytree.ca/quotes/Hermann.Goering.Quote.65D2 accessed 7.31.2020

5. Nina Renata Aron. This 1967 classroom experiment proved how easy it was for Americans to become Nazis. *Timeline* Jan 26 2017 https://timeline.com/this-1967-classroom-experiment-proved-how-easy-it-was-for-americans-to-become-nazis-ab63cedaf7dd

6. 'Snitches Get Rewards': Garcetti Issues New Rules for Construction Sites, Encourages Community to Report Safer At Home Violators. Mar 31 2020 https://losangeles.cbslocal.com/2020/03/31/coronavirus-los-angeles-eric-garcetti-snitches-get-rewards/ accessed 5.18.2020

7. Ariel Zilber. This is so un-American: Internet roasts Bill de Blasio after major urges New Yorkers to SNITCH on those not doing social distancing by snapping their photo and texting it to the city. *Daily Mail* April 19 2020 https://www.dailymail.co.uk/news/article-8233363/Mayor-Bill-Blasio-urges-New-Yorkers-SNITCH-not-doing-social-distancing.html accessed 5.18.2020

8. Milena Veselinovic and Laura Smith-Spark. UK Coronavirus adviser resigns after reports his lover visited during lockdown. *CNN World* May 6 2020 https://www.cnn.com/2020/05/05/uk/

neil-ferguson-imperial-coronavirus-sage-gbr-intl/index.htmlhttps://
www.cnn.com/2020/05/05/uk/neil-ferguson-imperial-coronavi-
rus-sage-gbr-intl/index.html accessed 9.1.2020

9. Tucker discovers Obama's 'essential' golf trip while under quarantine. *Fox News* April 30 2020 https://video.foxnews.com/
v/6153119056001#sp=show-clips accessed 9.1.2020

10. Vitale B. "Mike DeWine: Making Ohio Nice Again." *Columbus
Monthly* April 21 2020

11. Maeve Reston. "We must be at war with it: Mike DeWine remains collected
and directed in the face of the pandemic. *CNN* March 21 2020 https://www.
cnn.com/2020/03/21/politics/mike-dewine-coronavirus-response/index.
html accessed 9.1.2020

12. Jarrod Clay. Timeline of coronavirus in Ohio. *ABC6* Mar 31 2020
https://abc6onyourside.com/news/local/timeline-of-coronavi-
rus-in-ohio accessed 9.1.2020

13. GOVERNORS' POWER & AUTHORITY. National Governors
Association. https://www.nga.org/consulting-2/powers-and-authori-
ty/#role accessed 9.1.2020

14. Jack Windsor. Governor DeWine Suppresses Data Disproving
COVID-19 Policies. *Ohio Statehouse News* May 26 2020 https://
ohiostatehousenews.com/2020/05/governor-dewine-suppresses-da-
ta-disproving-covid-19-policies/ accessed 9.1.2020

15. IBID

16. COVID-19 Dashboard. Ohio Department of Health. https://coro-
navirus.ohio.gov/wps/portal/gov/covid-19/dashboards accessed
9.1.2020

17. Karen Kasler. Huge Percentage of Ohio COVID-19 Deaths Come From
Nursing Homes. *Statehouse News Bureau* May 21 2020 https://www.
statenews.org/post/huge-percentage-ohio-covid-19-deaths-come-nurs-
ing-homes accessed 9.1.2020

18. https://www.operationrescue.org/wp-content/uploads/2020/06/amy-action-ohio-apps.pdf accessed 9.1.2020

19. Cheryl Sullenger. Ohio Health Director's Mother Comes Forward to Set the Record Straight About Daughter's Troubled Past. *Operation Rescue* June 11 2020 https://www.operationrescue.org/archives/ohio-health-directors-mother-comes-forward-to-set-the-record-straight-about-daughters-troubled-past/ accessed 9.1.2020

20. https://vindyarchives.com/news/2019/aug/31/ohio-leaders-passion-comes-from-youngsto/

21. Cheryl Sullenger. Ohio Health Director's Mother Comes Forward to Set the Record Straight About Daughter's Troubled Past. *Operation Rescue* June 11 2020 https://www.operationrescue.org/archives/ohio-health-directors-mother-comes-forward-to-set-the-record-straight-about-daughters-troubled-past/ accessed 9.1.2020

22. Jacob Myers, Randy Ludlow. DeWine pledges to release number of coronavirus deaths I long-term care facilities. *Columbus Dispatch* Apr 20 2020 https://www.dispatch.com/news/20200420/dewine-pledges-to-release-number-of-coronavirus-deaths-in-long-term-care-facilities accessed 9.1.2020

23. Zanotti D. "Where Ohio's COVID-19 Strategy Went Wrong." American Policy roundtable. July 15 2020

24. Joe Hoft. CDC Reports that the China Coronavirus Mortality "Is Currently Below Pandemic Threshold." *Gateway Pundit* July 15 2020 https://www.thegatewaypundit.com/2020/07/cdc-reports-china-coronavirus-mortality-currently-pandemic-threshold/ accessed 9.1.2020

25. Ian Cross. Ohio reports 1316 new COVID-19 cases, 6 deaths; continued spike in hospitalizations. *ABC News5* July 15 2020 https://www.news5cleveland.com/news/continuing-coverage/coronavirus/ohio-reports-1-316-new-covid-19-cases-6-deaths-continued-spike-in-hospitalizations accessed 9.1.2020

26. Hospital Statistics by State. American Hospital Directory. https://www.ahd.com/state_statistics.html

27. https://www.news5cleveland.com/news/continuing-coverage/coronavirus/ohio-surpasses-70-000-coronavirus-cases-with-1-290-cases-28-deaths-reported-thursday accessed 7.16.2020

28. Cheryl Stephen. Parents of Terminal Child in ICU Kept Away; Family Please for Help. *Ohio Statehouse News* May 17 2020 https://ohiostatehousenews.com/2020/05/parents-of-terminal-child-in-icu-kept-away-family-pleads-for-help/ accessed 9.1.2020

29. Daniel Hampton. Cuomo Tells Out-Of-Work Protesters to Get Job As Essential Worker. *Patch* Apr 22 2020 https://patch.com/new-york/longisland/cuomo-tells-out-work-protesters-get-job-essential-worker accessed 8.2.2020

30. Marcy Oster. After hundreds at funeral, Cuomo tells ultra-Orthodox to halt large gatherings. *The Times of Israel* April 8, 2020 https://www.timesofisrael.com/new-york-governor-tells-ultra-orthodox-jews-to-halt-large-gatherings/ accessed 9.1.2020

31. Zimbardo P. *The Lucifer Effect* Random House New York New York 2007

32. COVID-19 Health Alerts. State of Alaska COVID-19 (CORONAVIRUS) INFORMATION https://covid19.alaska.gov/health-alerts/

33. IBID

34. Maggie Angst. "Coronavirus: Californians everywhere can now get their hair curt, nails done, and even a massage outside." *The Mercury News* July 20 2020 https://www.mercurynews.com/2020/07/20/newsom-expected-to-allow-california-salon-barber-shop-owners-to-work-outside/ accessed 8.1.2020

35. Eric A. Blair. What is a Meal? The New Rules or Bars and Restaurants in CA, NY Are Unbelievable. *Gateway Punfit* July 23

2020 https://www.thegatewaypundit.com/2020/07/meal-new-rules-bars-restaurants-ca-ny-unbelievable/ accessed 9.1.2020

36. COVID-19 and Sex. British Columbia Centre for Disease Control. http://www.bccdc.ca/health-info/diseases-conditions/covid-19/prevention-risks/covid-19-and-sex accessed 9.1.2020

37. Dan Lyman. California Restaurant Bans Face Masks. *NewsWars* Aug 1 2020 https://www.newswars.com/california-restaurant-bans-face-masks/ accessed 9.1.2020

38. Lisa Rantala. Ohio School district votes to violate state orders to give seniors delayed graduation." *Fox 28* July 10 2020 https://myfox28columbus.com/news/local/ohio-school-district-votes-to-violate-state-orders-to-give-seniors-delayed-graduation accessed 9.1.2020

39. Sherrie Peif. Weld County Attorney says health orders unenforceable; advises against playing waiver game. *Complete Colorado* June 26 2020 https://pagetwo.completecolorado.com/2020/06/26/weld-county-attorney-says-cdphe-orders-are-not-enforceable-weld-county-commissioners-advised-to-not-play-the-states-waiver-game-advised-to-not-play-waiver-game/?fbclid=IwAR1V-uSmJL2chs8le-dOhHMFPy4uhRz8xihr-kM-qxGjMKIkLMd5oAb0ZxRw accessed 9.1.2020

40. The Rise of the Anti-Lockdown Sheriffs. The Marshall Project. https://www.themarshallproject.org/2020/05/18/the-rise-of-the-anti-lockdown-sheriffs accessed 8.1.2020

41. Bowen Xiao. Sheriffs: Lockdown Measures Are Unconstitutional, We Won't Enforce Them. *Epoch Times* May 27 2020 https://www.theepochtimes.com/sheriffs-lockdown-measures-are-unconstitutional-we-wont-enforce-them_3366351.html accessed 9.1.2020

42. 'I am not the mask police': Butler county sheriff says he won't enforce mandatory masks. *WLWY5* July 7 2020 https://www.wlwt.

com/article/i-am-not-the-mask-police-butler-county-sheriff-says-he-wont-enforce-mandatory-masks/33237101 accessed 9.1.2020

43. Georgia governor explicitly voids mask orders in 15 localities. *News4Jax* https://www.news4jax.com/news/georgia/2020/07/16/georgia-governor-explicitly-voids-mask-orders-in-15-locali-ties/?utm_source=facebook&utm_medium=social&utm_cam-paign=snd&utm_content=wjxt4 accessed 8.1.2020

44. Phil Helsel. Mississippi church fighting coronavirus restrictions burned to the ground. *NBC News* May 22 2010 https://www.nbcnews.com/news/us-news/mississippi-church-fighting-coronavi-rus-restrictions-burned-ground-n1212646 accessed 9.1.2020

45. Yael Halon. New Jersey gym owner defiant after arrest for violating stay-at-home order: 'I'm not afraid of tyrants. *FOX News* https://www.foxnews.com/media/new-jersey-gym-owner-not-afraid-tyrants accessed 8.1.2020

46. Bobby Caina Calvan. Florida pair arrested for breaking COVID-19 quarantine order. *AP* July 30 2020 https://apnews.com/d813ace-59029914702da962af92a9c2a accessed 8.1.2020

47. Faith King. Hardin County couple on house arrest after not signing positive COVID-19 self-isolation order. *WAVE News* July 17 2020 https://www.wave3.com/2020/07/17/hardin-county-couple-house-arrest-after-not-signing-positive-covid-self-isolation-order/ accessed 8.1.2020

48. Erin Robinson. Spokane County man with COVID-19 taken to jail for refusing to self-isolate. *KXLY.com* July 2 2020 https://www.kxly.com/spokane-county-man-with-covid-19-taken-to-jail-for-refusing-to-self-isolate/ accessed 9.1.2020

49. Arkansas National Guard transporting COVID-19 pa-tients to isolation facility. *Fox 13* July 15 2020 https://www.fox13memphis.com/news/local/

arkansas-national-guard-transporting-covid-19-patients-isolation-facility/M6QOEDIC4JEGLKO7L2AN4BVYDQ/ accessed 8.1.2020

50. Kevin Daly and George Miller. Forced County COVID-19 Lockdown of Ventura County Apt Building at 137 S. Palm St. *Citizens Journal* July 20 2020 https://www.citizensjournal.us/forced-county-covid-19-lockdown-of-ventura-apt-building-137-s-palm-st/ accessed 9.1.2020

51. ABC7.com staff. Security Guard suffers broken arm after confrontation over masks inside Van Nuys Target. *Eyewitness News* May 11 2020 https://abc7.com/van-nuys-target-mask-fight-surveillance-video-face-requirement/6172453/ accessed 9.1.2020

52. Trina Orlando. McHenry Woman arrested After Fight About Face Masks at Home Depot. *NBC5Chicago* July 6 2020 https://www.nbcchicago.com/news/local/mchenry-woman-arrested-after-fight-about-facial-coverings-at-home-depot/2300635/#:~:text=NBCUniversal%2C%20Inc.,at%20Home%20Depot%2C%20police%20said. Accessed 9.1.2020

53. Guardian Staff. Officer fatally shoots Michigan man after mask dispute leads to stabbing. *The Guardian* July 14 2020 https://www.theguardian.com/us-news/2020/jul/14/michigan-face-mask-stabbing-shooting accessed 9.1.2020

54. OSHA Bulletin. General Respiratory Protection Guidance for Employers and Workers. United States Department of Labor. https://www.osha.gov/dts/shib/respiratory_protection_bulletin_2011.html accessed 9.1.2020

55. Medical Clearance and Fit-test Procedures for M95 Masks-COVID-19. OSHA Review. https://oshareview.com/2020/05/medical-clearance-and-fit-test-procedures-for-n95-masks-covid-19/ accessed 9.1.2020

56. "Medical masks are a tool that can be used to prevent the spread of respiratory infection." *JAMA* 2020 Apr;323(15):1517-1518

57. Kerry Breen. Can you resuse a disposable mask? Yes, if you follow these steps. *Yahoo News* July 14 2020 https://news.yahoo.com/reuse-disposable-mask-yes-steps-192223361.html accessed 9.1.2020

58. Preventing transmission of pandemic influenza and other viral respiratory diseases: personal protective equipment for health-care workers: update 2010. Institute of Medicine (US) Committee on Personal Protective Equipment for Healthcare Personnel to Prevent Transmission of Pandemic Influenza and Other Viral Respiratory Infections: Current Research Issues; Editors: Elaine L. Larson and Catharyn T. Liverman Washington: The National Academies Press; 2011.

59. Oberg T, Brosseau LM. "Surgical mask filter and fit performance." *Am J Infect Control* 2008 May;36(4):276-282

60. MacIntyre CR, Cauchemez S, Dwyer DE et al. "Face mask use and control of respiratory virus transmission in households." *Emerg Infect Dis* 2009 Feb;15(2):233-241

61. Cowling BJ, Chan KH, Fang VJ et al. "Facemasks and hand hygiene to prevent influenza transmission in households: a cluster random-ized trial." *Ann Intern Med* 2009 Oct;15(7):437-446

62. Yassi A, Bryce E. Protecting the Faces of Health Care Workers. Occupational Health and Safety Agency for Healthcare in BC, Final Report, April 2004. http://www.phsa.ca/Documents/Occupational-Health-Safety/ReportProtectingtheFacesofHealthcareWorkers.pdf accessed 9.1.2020

63. Bin-Reza F, Chavarrias VL, Nicoll A, Chamberland ME. "The use of masks and respirators to prevent transmission of influenza: a system-atic review of the scientific evidence." *Influenza Other Respir Viruses* 2012 Jul;6(4):257-267

64. Desai AN, Mehrotra P. "Medical Masks." *JAMA* 2020 Mar;323(15):1517-1518

65. Face Masks and Surgical Masks for COVID-19: Manufacturing, Purchasing, Importing, and Donating Masks During the Public Health Emergency. U.S. Food and Drug Administration. https://www.fda.gov/medical-devices/personal-protective-equipment-infection-control/face-masks-and-surgical-masks-covid-19-manufacturing-purchasing-importing-and-donating-masks-during accessed 9.1.2020

66. MccIntyre CR, Seale H, Dung TC et al. "A cluster randomised trial of cloth masks compared with medical masks in healthcare workers." *BMJ Open* 2015 Mar;5(4):e006577

67. Xiao J, Shiu EYC, Gao H et al. "Nonpharmaceutical Measures for Pandemic Influenza in Nonhealthcare Settings – Personal Protective and Environmental Measures." *Emerging Infectious Diseases* 2020 May;26(5):967-975

68. Jonathon Lapook. CORONAVIRUS; HOW U.S. HOSPITALS ARE PREARING FOR COVID-19, AND WHAT LEADING HEALTH OFFICIALS SAY ABOUT THE VIRUS. *60 Minutes* Mar 8 2020 https://www.cbsnews.com/news/coronavirus-containment-dr-jon-lapook-60-minutes-2020-03-08/

69. Grace Panetta. Fauci says he doesn't regret telling Americans not to wear masks at the beginning of the pandemic. *Business Insider* Jul 16 2020 https://www.businessinsider.com/fauci-doesnt-regret-advising-against-masks-early-in-pandemic-2020-7 accessed 9.1.2020

70. Charles Spiering. Dr. Anthony Fauci Opposes Controlled Study on Effectiveness of Masks. *Breitbart* July 16 2020 https://www.breitbart.com/politics/2020/07/16/dr-anthony-fauci-opposes-controlled-study-effectiveness-masks/ accessed 9.1.2020

71. Caterina Andreano. Dr. Fauci: Wear goggles or eye shields to prevent spread of COVID-19; flu vaccine a must. *ABC News* Jul 29 2020 https://abcnews.go.com/US/dr-fauci-wear-goggles-eye-shields-prevent-spread/story?id=72059055#:~:text=Dr.%20Anthony%20

Fauci%20suggested%20Wednesday,spreading%20or%20catch-ing%20COVID%2D19. Accessed 9.1.2020

72. Michael Martiromo. Maskme: Suffering from acne or breakouts under your mask? Here's what to do. *Fox News* May 26 2020 https://www.foxnews.com/lifestyle/maskne-acne-breakouts-under-mask-what-to-do accessed 9.1.2020

73. Lim ECH, Seet RCS, Lee K-H, Wilder-Smith EPV, Chuah BYS, Ong BKC. "Headaches and the N95 Face-Mask Amongst Healthcare Providers." *Acta Neurol Scand* 2006 Mar;113(3):199-202

74. Ong JJY, Bharatendu C, Goh Y et al. "Headaches Associated With Personal Protective Equipment – A Cross-Sectional Study Among Frontline Healthcare Workers During COVID-19." *Headache* 2020 May;60(5):864-877

75. Robert Gearty. NJ police say 'excessive wearing' of coronavirus mas contributed to driver passing out, crashing car. *Fox News* April 25 2020 https://www.foxnews.com/us/nj-police-say-excessive-wearing-of-n95-coronavirus-mask-contributed-to-woman-passing-out-crash-ing-car accessed 9.1.2020

76. Beder A, Buyukkocak U, Sabuncuoglu A, Jeskil ZA, Keskil S. "Preliminary report on surgical mask induced deoxygenation during major surgery." *Neurocirugia (Astur)* 2008 Apr;19(2):121-6

77. Fletcher SJ, Clark M, Stanley PJ. "Carbon dioxide re-breathing with close fitting face respirator masks" *Anaesthesia* 2006 Sep;6(9):910

78. Kyung SY, Kim Y, Hwang H, Park JW, Jeong SH. "Risks of N95 Face Mask Use in Subjects With COPD." *Respir Care* 2020 May;65(5):658-664

79. https://www.dailymail.co.uk/news/article-8283965/Two-Chinese-boys-drop-dead-run-PE-lessons-wearing-face-masks.html accessed 7.18.2020

80. Emilia Jiang. "Two Chinese boys drop dead during PE lessons while wearing face masks amid concerns over students' fitness following three months of school closure. *Daily Mail* May 4 2020 https://www.dailymail.co.uk/news/article-8311179/Joggers-lung-collapses-ran-2-5-miles-wearing-face-mask.html?ito=facebook_share_article-facebook_preferred-top&fbclid=IwAR0kieVZJ9qUeNir6ELHbdys-4KoOJqfk6Wsz-RknRDXWrQZCpBRr--br2A0 accessed 7.18.2020

81. Angela Betsaida B. Laguipo. "Wearing masks may increase your risk of coronavirus infection, expert says." *News Medical Life Sciences* Mar 15 2020 https://www.news-medical.net/news/20200315/Reusing-masks-may-increase-your-risk-of-coronavirus-infection-expert-says.aspx accessed 9.1.2020

82. Nisha Kurani, Karen Politz, Dustin Cotliar, Nicolas Shanosky, Cynthia Cox. COVID-19 Test Prices and Payment Policy. Health System Tracker. https://www.healthsystemtracker.org/brief/covid-19-test-prices-and-payment-policy/ accessed 9.1.2020

83. COVIDView Summary ending on April 18 2020. Centers for Disease Control and Prevention. https://www.cdc.gov/coronavirus/2019-ncov/covid-data/covidview/past-reports/04242020.html accessed 8.1.2020

84. COVIDView. A Weekly Surveillane Summary of U.SS. COVID-19 Activity. Centers for Disease Control and Prevention. https://www.cdc.gov/coronavirus/2019-ncov/covid-data/covidview/index.html accessed 8.1.2020

85. Standardized surveillance case definition and national notification for 2019 novel coronavirus disease (COVID-10). Council of State and Territorial Epidemiologists. https://cdn.ymaws.com/www.cste.org/resource/resmgr/2020ps/Interim-20-ID-01_COVID-19.pdf accessed 8.1.2020

86. Data management for Assigning and Managing Investigations. Centers for Disease Control and Prevention. https://www.cdc.gov/

coronavirus/2019-ncov/php/contact-tracing/contact-tracing-plan/
data-management.html 8.1.2020

87. Petr Svab. At Least 22 States count 'Probable' COVID Cases or
    Deaths in Totals. *Epoch Times* July 21 2020 https://www.theepoch-
    times.com/at-least-22-states-count-probable-covid-cases-or-deaths-
    in-their-totals_3432988.html accessed 9.1.2020

88. Dashboard of Public Health Indicators. Massachusetts Department
    of Public Health COVID-19 Dashboard. https://www.mass.gov/doc/
    covid-19-dashboard-august-1-2020/download accessed 8.1.2020

89. HHS to Begin Distributing $10 Billion in Additional Funding
    to Hospitals in High Impact CoVID-19 Areas. July 17 2020
    https://www.hhs.gov/about/news/2020/07/17/hhs-begin-dis-
    tributing-10-billion-additional-funding-hospitals-high-im-
    pact-covid-19-areas.html?fbclid=IwAR02xgfUNCAcR1M7Tr7M-
    cHmZPu5MD7YvtG_5-UCtgL983DA7dAhior5tEvQ accessed
    8.1.2020

90. Edwin Mora. CDC Chief Agrees There's Perverse Economic
    Incentive for Hospitals to Inflate Coronavirus Deaths. *Breitbart
    News* July 31 2020 https://www.breitbart.com/politics/2020/07/31/
    cdc-chief-agrees-theres-perverse-economic-incentive-for-hospitals-
    to-inflate-coronavirus-deaths/ accessed 8.1.2020

91. CT State Lab Finds 90 Positive COVID-19 Test Results Were
    False. https://www.nbcconnecticut.com/news/coronavirus/
    state-says-90-positive-covid-9-test-results-were-false/2304893/

92. Laura Gutschke. Texas officials investigating false-positive
    COVID-19 tests at area nursing homes. *Abilene Reporter News*
    Jun 4 2020 https://www.reporternews.com/story/news/2020/06/04/
    state-investigating-false-positive-covid-19-results-snyder-east-
    land-cisco-nursing-homes/3145227001/ accessed 9.1. 2020

93. Robert Kraychik. Fox 35 Investigation Reveals Inflated Florida
    Coronavirus Numbers. *Breitbart* Jul 14 2020 https://www.breitbart.

com/health/2020/07/14/fox-35-investigation-reveals-inflated-flori-da-covid-19-numbers/ accessed 9.1.2020

94. Coronavirus (COVID-19) Update: Facilitating Diagnostic Test Availability for Asymptomatic Testing and Sample Pooling. U.S. Food and Drug Administration. https://www.fda.gov/news-events/press-announcements/coronavirus-covid-19-update-facilitating-diagnostic-test-availability-asymptomatic-testing-and accessed 9.1.2020

INFLATED CASES AND DEATHS

AS YOU'VE LEARNED, the number of cases and deaths did not warrant the declaration of a pandemic and certainly did not warrant drastic interventions in the U.S. such as shelter-in-place, closure of most businesses, sending children home from school, and loss of personal freedoms. But even the low numbers were almost certainly inflated.

Inaccurate Testing

From the beginning, COVID-19 testing in the U.S. has been flawed. While the World Health Organization had developed testing specifications for COVID-19 by January 2020, the CDC decided to develop its own test, which was ready by early February. The test was manufactured and distributed by the CDC to health centers throughout the U.S., and within a few days, the tests were found to be inaccurate. In response the FDA insisted that hospitals, academic centers and private companies *should not develop their own tests.* When the agency finally lifted the ban on test development at the end of February, there was a rush to get tests ready for market. Although the FDA provided no standards for how COVID-19 was to be detected. This meant all test makers could decide what standard to use.

Over 100 companies are currently producing tests for COVID-19, and these tests were approved by the FDA under emergency authorization with minimal validation. The test makers only had to show that the tests performed well in test tubes and no real-world demonstration of clinical viability was required.[1] Each vendor established its own and as-yet-unmeasured accuracy. The variations are myriad, with some tests able to detect as few as 100 copies of a viral gene while others require 400 copies for detection.[2] Additionally, most will show positive results for as long as 6 months, while the actual time a person is contagious is only a few days.

Several issues were never addressed. One is the potential cross-reactivity with other viruses. Another is that the presence of coronavirus is likely to remain for several months after the infectious period has passed. This means the tests are useless for determining who should be quarantined. Yet another is the risk of cross contamination, particularly when testing large numbers of people in crowded settings. Even the tiniest amount of cross contamination can lead to a false positive result, which means people who have never been exposed to COVID-19 could be subjected to unwarranted quarantines.

The accuracy of tests is important since numbers of "cases" is the metric used to determine business closures, event cancellations, lockdowns, withdrawal of civil rights and liberties, whether people can congregate, and if the useless masks are required.

There are two primary processes used to test for the coronavirus. The first method requires a sample of mucus from a person's nose or throat and then attempting to replicate the RNA through a Polymerase Chain Reaction (PCR) machine. The second is through the antibody test, a blood test that is supposed to determine not if one is infected, but if they have ever been infected. Both tests are flawed.

Biochemist Kary Mullis is the inventor of the PCR test and won the Nobel Prize in chemistry for his invention in 1993. Mullis stated in 2013 that PCR was never designed to diagnose disease. The test finds very small segments of a nucleic acid which are components of a virus. According to Mullis, having an actual infection is quite different than testing positive with PCR. According to Mullis, PCR is best used in medical laboratories and for research purposes.

Dr. David Rasnick, also a biochemist and founder of a lab called Viral Forensics, agrees.

"You have to have a whopping amount of any organism to cause symptoms. Huge amounts of it. You don't start with testing; you start with listening to the lungs. I'm skeptical that a PCR test is ever true. It's a great scientific research tool. It's a horrible tool for clinical medicine. 30% of your infected cells have been killed before you show symptoms. By the time you show symptoms...the dead cells are generating the symptoms."

When asked about having a COVID-19 test he stated, "Don't do it, I say, when people ask me. No healthy person should be tested. It means nothing but it can destroy your life, make you absolutely miserable." He went on to say, "Every time somebody takes a swab, a tissue sample of their DNA, it goes into a government database. It's to track us. They're not just looking for the virus. Please put that in your article."[3]

In fact, PCR testing was already shown to be wildly inaccurate almost 15 years ago.

In 2006, massive PCR testing was performed at the Dartmouth Hitchcock Medical Center when it was thought that the medical center was experiencing an epidemic of whooping cough. Almost 1000 healthcare workers were furloughed until their test results were returned. Over 140 employees were told that they had whooping cough, and thousands of others who tested positive were given antibiotics and/or a vaccine for whooping cough.

Almost eight months later, employees received an email from the hospital administration which stated that the entire episode was due to PCR testing error. Not even one case of whooping cough was confirmed with a more reliable follow-up test, and it was determined that the employees just had a common cold, not whooping cough.[4]

Apparently, this history was ignored as incompetent health officials like Mr. Fauci decided that ginning up cases was more important than following the science. Thus, a test that the developer said was not useful for diagnosis and that had been previously shown to be inaccurate 100% of the time was recommended for COVID-19.

A recent meta-analysis published in the *British Medical Journal* looked at the accuracy of PCR testing specifically for COVID-19. The researchers reported that while no test is 100% accurate, the sensitivity and specificity of a test is evaluated by comparison with a gold standard, and there is no gold standard for COVID-19. One of the reasons is that it is impossible to know the false positive rate without having tested people who don't have the virus along with people who do, and this was never done.

The analysis showed that the false negative rate ranges between 2% and 29%. Accuracy of viral RNA swabs was highly variable. In one study, sensitivity was 93% for bronchoalveolar lavage, 72% for sputum, 63% for nasal swab, and only 32% for throat swabs. The researchers stated that results vary for many reasons including stage of disease.[5] This analysis was published in May, long after Mr. Fauci and his accomplices had succeeded in creating a false pandemic, in part by insisting that more and more people should be tested.

Fortunately, many people are far more diligent than Fauci in checking out facts.

Investigators from *OffGuardian* contacted the authors of four papers published in early 2020 in which researchers claimed that they had discovered a new coronavirus. The investigators asked for proof that electron micrographs showed purified virus and all four groups replied that they did not.

Here are the verbatim responses from the four groups:

"The image is the virus budding from an infected cell. It is not purified virus."

"We could not estimate the degree of purification because we do not purify and concentrate the virus cultured in cells."

"[We show] an image of sedimented virus particles, not purified ones."

"We did not obtain an electron micrograph showing the degree of purification."

The investigators also contacted virologist Charles Calisher and asked if he knew of any research group that had isolated and purified SARS-COV-2 and he replied that he did not. They concluded at this time no one knows whether the RNA gene sequences used in the in vitro trials and which were used to calibrate the tests came from SARS-CoV-2.[6]

All of this may explain why some of the testing results from around the world have been so difficult to understand or explain. For example, testing in Guangdong province in China showed that 10% of people who recovered from COVID-19 tested negative and then tested positive again.[7] Twenty-nine patients tested in Wuhan tested negative, then positive, and then the results were "dubious."[8]

According to Wang Chen, president of the Chinese Academy of Medical Sciences, PCR tests are only 30-50% accurate.[9]

The FDA agrees. A statement in its online instruction manual for PCR testing includes these statements:

Detection of viral RNA may not indicate the presence of infectious virus or that 2019-nCoV is the causative agent for clinical symptoms."

This test cannot rule out diseases caused by other bacterial or viral pathogens."[10]

The FDA's online emergency use authorization includes this statement:

"positive results [...] do not rule out bacterial infection or co-infection with other viruses. The agent detected may not be the definite cause of disease."[11]

In fact, the manufacturer's instruction manual for one PCR test includes these statements:

These assays are not intended for use as an aid in the diagnosis of coronavirus infection"

For research use only. Not for use in diagnostic procedures."[12]

The bottom line is that this test is useless for diagnosing COVID-19. If the error rate is only 5% this could mean that the number of cases worldwide is off by millions. But the error rate is most likely much higher, which means that the world's population is suffering due to a made-up pandemic.

There Are Other Serious Issues

Some county and state health departments state that the counts for coronavirus are typically reported via a primary care physician or pulmonologist.[13] Most likely neither of these provider types has an expensive PCR machine at their disposal. Thus, it would appear as though the virus is being diagnosed by physicians the same way they would diagnose any common cold or flu, which is by physical examination and observation of symptoms. The symptoms of COVID-19 are like those of influenza in many ways.

Several Governors in the U.S. requested billions of dollars in federal aid to "assist with the impact of the coronavirus," the amount of which was based on the infection rate. Collectively, they requested a total of $500 billion.[14] At this time there is no accountability for exactly how this aid was spent. It is interesting that the states with the worst per capita debt (such as California and New York) have requested the most money.[15] Coincidence? Perhaps not. Naturally, it could make sense to report a higher rate of infection in order to receive a larger piece of the stimulus.

There have been numerous problems with the testing procedures, some political, some scientific. The CDC went against the guidance of the World Health Organization (WHO).[16] The irony, of course, is one corrupt organization ignoring the guidelines of another corrupt organization. Ultimately, the missteps that occurred regarding testing were massive. On April 20, 2020, it was reported that the tests the CDC was using were contaminated with the coronavirus itself.[17] There was no way to know the number of false negatives and false positives.

The Food and Drug Administration (FDA) sent representatives to the CDC and found the primary culprit to be poor laboratory practices. The CDC offered no defense for its decisions.

Testing was not much better in other parts of the world. For example, Spain and the Czech Republic spent millions on a test purchased from a Chinese company called "Shenzhen Bioeasy Technology" and later found that the tests were only 30% accurate. Gordon Chang, who has covered Chinese economics and policy for decades stated "It [China] creates the poison and then sells the cure to it."[18] How purposeful was this? We will never know, although China had an incentive to keep the world frightened and shut down, both to gain economic advantage and to distract the world while it engaged in practices condemned by many countries.

Even if the test kits are not faulty, more false negatives can result from the swabbing method used to collect samples. The tests typically require a swab to be inserted into the nasal passage. This is a common method used in the "drive-thru" testing sites set up in many cities. In order to be properly detected, the swab must be inserted deep into the nasal passage, causing considerable discomfort. Many of those performing the tests were either not properly trained or tended to withdraw the swab early when the patient exhibited discomfort or resistance.

Dr. Michael Pintella, Director of the State Hygienic Lab in Iowa, stated "Tests involve a multi-step process and each step might lead to a false negative result for any number of reasons, including a poorly collected specimen, a delay in transport of the specimen to the lab, not storing or transporting specimens at the appropriate temperature, problems encountered during testing extraction, analysis errors and more."[19] In the same news release Dr. Austin Baeth, who was very outspoken about wanting to administer a state lockdown for Iowa, admitted that the tests only have a 63% accuracy rate.

The other common method for testing is the antibody test, which uses a blood sample. The problem with this test is that it does not determine if one has the virus, rather if one has had it before. This is also problematic, as there are many false positives due to detecting antibodies created from other coronaviruses (such as the common cold).[20] The methodology is flawed as well. According to a report released in early May, the FDA had to tighten restrictions on the hundreds of companies that were profiting from selling fraudulent testing kits.[21] Some of these kits were even being advertised as "do it yourself from home" products. It is widely believed that there are many false negatives arising from these kits as well.

To make matters worse, the CDC had been reporting positive test results from a combination of both the PCR test and the antibody test. Ashish Jha, the K.T. Li Professor of Global Health at Harvard University said, "You've got to be kidding me. How could the CDC make that mistake? This is a mess." [22] He further went on to say that mixing the results of the two tests muddies the water. One test is like looking in the rearview mirror and the other just says if one is infected now. He also stated that because of this, the actual amount of cases is and was much higher than reported.

Testing in Tanzania:
Apparently Fruit Can Test Positive

The head of Tanzania's health laboratory in charge of coronavirus was suspended after President John Magufuli of Tanzania had a security detail obtain random samples of Pawpaw, jackfruit, and animals which tested positive for COVID-19.

Samples of fruit were taken from inside the fruit – therefore positive results could not be from someone touching the fruit. The samples were given names and sent to the laboratory.

Here were the results:

- Sample of car oil named Jabil Hamza, 30 years old, male - negative
- Sample from Jackfruit named Sarah Samuel 45 years old, female - inconclusive test results
- Sample of liquid from Pawpaw named Elizabeth Anne 26 years old female - positive
- Samples from Kware (type of bird) – positive
- Samples from rabbit – undetermined
- Goat – positive
- Sheep – negative

Magufuli said that this means the Pawpaw named Elizabeth must be placed in isolation, goats should be in isolation, and Jackfruit named Sara should be in isolation. But, he reported, the Pawpaw is not dying it's just getting ripe. Magufuli says, "a dirty game is being played with these tests," reported that the tests were imported, and said the WHO should do something about this. He told Reuters that this indicates that some people are testing positive who not have the disease.

The Centers for Disease Control and Prevention says there is no way that fruit can contract COVID-19

As of May 6, 2020, there were 480 cases and 17 deaths in Tanzania, and there was no way to know if the goats, sheep, bird, Pawpaw and jackfruit were included in the count.[23] [24]

But You Must Have COVID-19! You Must!

NBC referred to Dr. Joseph Fair as "…Today's most knowledge-able expert on the coronavirus outbreak." Dr. Fair reported that he was recently diagnosed with COVID-19, and tweeted that he was hospitalized with it.

According to Dr. Fair, he flew home from New York City to New Orleans wearing a mask and gloves, wiped everything down but says he must have contracted it through his eyes. He said that his symptoms were not classic symptoms, but when he developed shortness of breath, he called an ambulance and was admitted to Tulane Medical Center. He had four COVID tests and they were all negative, but he knows he had it and his doctors confirmed that this was the case.

It seems that anyone determined to have COVID-19 will have it – testing does not matter. Apparently, nor does wearing masks and gloves and wiping things down.[25]

And If All Else Fails, Use "Medical Intuition"

An article in Medscape posted May 16, 2020 describes a pa-tient who arrived at UC San Diego Health medical center with classic COVID-19 symptoms – a history of cough, pneumonia, severe respiratory distress – and required immediate intubation. The patient's back of the throat was swabbed twice and both times was negative for COVID-19. "The two negative tests didn't convince anybody," said Davey Smith, MD, a virologist and chief of the division of infectious diseases and global public health at UC San Diego School of Medicine. It was only on the third test, when they sampled fluid from a bronchial wash, that they were able to find the virus.[26] The article was titled "Don't Discount Medical Intuition."

The article went on to say that this is not an isolated incident because there are limitations to current tests and that clinicians report false negative rates as high as 30%. The FDA issued an alert warning of false negatives with Abbott Labs' ID NOW rapid test, one of the most used.[27]

The authors also cited data in *Annals of Internal Medicine* showing that test accuracy depends on when the person is tested because the false negative results vary during the course of the disease. According to this study, on the day symptoms appear, the false negative rate was 38%; it dropped to 20% on the third day and increased to 66% two weeks later.[28]

According to Stephen Rawlings, MD PhD, infectious disease fellow at UC San Diego Center for AIDS research, one of the problems is that there is nothing to compare current tests to. He says, "To truly determine false negatives, you need a gold standard test, which is essentially as close to perfect as we can get," Rawlings said. "But there just isn't one yet for coronavirus."

Colin West MD PhD at Mayo Clinic says that the studies that have looked at accuracy of tests currently used have been "filled with flaws," one of which is that the sensitivity estimates are based on testing people who the researchers already knew had COVID-19. This results in significant bias. He says that without control groups of blinded testing it's impossible to determine the magnitude of the inaccuracy.[29]

The results of an analysis of five studies that included 957 patients and that had yet to be peer-reviewed concluded that "The certainty of the evidence was judged as very low, due to the risk of bias, indirectness, and inconsistency issues. Conclusions: The collected evidence has several limitations, including risk of bias issues, high heterogeneity, and concerns about its applicability."[30]

Other Countries Inflated Numbers Too

Public health officials in the UK have inflated the number of cases by counting each test twice. When diagnostic tests were used that involved taking both saliva and nasal samples from the same patient, the results were counted as two separate tests. This led to inflated case numbers. Both the Department of Health and Social Care and Public Health England acknowledged that they had engaged in this practice.

This is not the only instance in which the UK government was caught inflating data.

In April, public health authorities included thousands of home tests which had been mailed out but not completed in order to make it look like the goal of 100,000 tests was being met.

Apparently using fake numbers to promote a fake pandemic is not limited to the U.S.[31]

The CDC's Strange Definition of a "Case"

As you have seen, the tests were definitely flawed. But the CDC's definition of a "case" did not require any testing at all. The CDC listed over one dozen ways in which a person could be diagnosed with COVID-19.

Here are excerpts from the CDC's "2020 Interim Case Definition"[32] (verbatim with commentary)

Clinical Criteria

At least two of the following symptoms: fever (measured or subjective), chills, rigors, myalgia, headache, sore throat, new olfactory and taste disorder(s)

> **OR**

At least one of the following symptoms: cough, shortness of breath, or difficulty breathing

> **OR**

Severe respiratory illness with at least one of the following:
Clinical or radiographic evidence of pneumonia **OR**
Acute respiratory distress syndrome

> **AND**

No alternative more likely diagnosis

Commentary on "Clinical Criteria"

Note that fever can be "subjective."

Headache, sore throat and cough can be symptoms of many things, including the common cold.

"New olfactory and taste disorders" An article published in the *Lancet* referred to COVID testing as "inadequate" and suggests that new symptom profiles be developed to help identify those who should be quarantined.

It suggests that loss of taste and smell are highly predictive of COVID-19 and anyone experiencing these symptoms should self-isolate.[33]

In fact, there are many causes of loss of taste and smell. These include:

- Aging especially after age 60
- Allergies
- Nasal and sinus problems like sinusitis or nasal polyps
- Medications including beta blockers and ACE inhibitors
- Dental problems
- Cigarette smoking
- Head or facial injury
- Alzheimer's disease
- Parkinson's disease
- Common cold or other viral infections (40%)[34]

In fact, as much as 20% of the general population has a prolonged smell disorder.[35]

There are many problems with the *Lancet* article. The basis for the recommendation to use taste and smell as a diagnostic tool is data collected from patients using an online app. Almost 60% of 579 people who reported testing positive said they had lost their sense of smell and taste; but almost 18% of the 1123 who tested negative also reported loss of taste and smell.[36]

The researchers acknowledge many limitations which include that these symptoms are non-specific and lack predictive power, and their report relied on self-reported information, which is generally unreliable. Yet, they write, "We believe that having added loss of smell and taste to the list of COVID-19 symptoms is of great value as it will help trace almost 16% of cases that otherwise would have been missed. Loss of smell and taste, together with fever or cough, should now enable us to identify 87.5% of symptomatic COVID-19 cases, although this is likely to be less in the early phases of the infection." This conclusion is hard to fathom in consideration of the facts, although facts have not seemed to matter much these days.

Here's a much more realistic assessment from Eric Holbrook, director of rhinology at Massachusetts Eye and Ear: "Physicians are collecting data so quickly, but a lot of it is subjective data. I haven't seen a careful study that looks at when patients get the diagnosis, and how severe it is, and how long the smell loss lasts."[37]

Laboratory Criteria

Laboratory evidence using a method approved or authorized by the U.S. Food and Drug Administration (FDA) or designated authority:

Confirmatory laboratory evidence:

- Detection of severe acute respiratory syndrome coronavirus 2 ribonucleic acid (SARS-CoV-2 RNA) in a clinical specimen using a molecular amplification detection test

Presumptive laboratory evidence:

- Detection of specific antigen in a clinical specimen
- Detection of specific antibody in serum, plasma, or whole blood indicative of a new or recent infection*

  Serologic methods for diagnosis are currently being defined

Commentary on Laboratory Criteria:

Note that these are the tests we proved were inaccurate, and that the CDC admits that the serological methods for diagnosis are currently being defined, but they are ok to use for purposes of diagnosis now.

Epidemiologic Linkage

One or more of the following exposures in the 14 days before onset of symptoms:

- Close contact** with a confirmed or probable case of COVID-19 disease; **OR**
- Close contact** with a person with:
  - clinically compatible illness **AND**
  - linkage to a confirmed case of COVID-19 disease.
- Travel to or residence in an area with sustained, ongoing community transmission of SARS-CoV-2.
- Member of a risk cohort as defined by public health authorities during an outbreak.

**Close contact is defined as being within 6 feet for at least a period of 10 minutes to 30 minutes or more depending upon the exposure. In healthcare settings, this may be defined as exposures of greater than a few minutes or more. Data are insufficient to precisely define the duration of exposure that constitutes prolonged exposure and thus a close contact.

Commentary on Epidemiologic Linkage:

- A person who has been within 6 feet of someone for 10 minutes who may have but is not confirmed to have COVID-19 is now considered a case
- A person who has been within 6 feet of a person who has a headache or a sore throat, or has changes in smell or taste is now considered a case
- A person who has been in contact with a person who is linked to a person with COVID-19 is now a case.
- Travel to an area in which there are COVID-19 cases qualifies a person as a case.
- Being a member of a "risk cohort" also qualifies a person as a case. There are no examples, but a statement that health authorities can just name a group as a risk category.
- The CDC acknowledges that it is not known the length of exposure required to cause a problem but uses this metric anyway.

Criteria to Distinguish a New Case from an Existing Case

Not applicable (N/A) until more virologic data are available.

Commentary on Criteria to Distinguish a New Case from an Existing Case:

The CDC does not know how to determine a new from an existing case, but when trying to gin up cases, what difference could this make?

Ginning Up the Death Rate

According the CDC's document titled "Guidance for Certifying Deaths Due to Coronavirus Disease 2019 (COVID–19)":[38]

"In cases where a definite diagnosis of COVID cannot be made but is suspected or likely (e.g. the circumstances are compelling with a reasonable degree of certainty) it is acceptable to report COVID-19 on a death certificate as 'probable' or 'presumed.'"

In other words, when in doubt, classify any death possible as COVID-19, which will serve to inflate the numbers to make it look like the projections are right and keep the hoax alive.

The National Vital Statistics System issued an alert on March 24, 2020 regarding a new ICD code for COVID-19 deaths. According to this document:

The WHO has provided a second code, **U07.2**, for clinical or epidemiological diagnosis of COVID-19 where a laboratory confirmation is inconclusive or not available.

Will COVID-19 be the underlying cause?

The underlying cause depends upon what and where conditions are reported on the death certificate. However, the rules for coding and selection of the underlying cause of death are expected to result in COVID-19 being the underlying cause more often than not.

Should "COVID-19" be reported on the death certificate only with a confirmed test?

COVID-19 should be reported on the death certificate for all decedents where the disease caused **or is assumed to have caused or contributed to death**.[39]

Again, specific instructions to list the cause of death as COVID-19 as much as possible.

Dr. Deborah Birx, a member of the White House task force, confirmed this. She announced during a press briefing on Tuesday April 7, 2020 that the deaths of all patients who died with coronavirus, even if the cause of death was not due to COVID-19, should list COVID-19 as cause of death on the death certificate. She acknowledged that other countries do not do this. "There are other countries that if you had a pre-existing condition, and let's say the virus caused you to go to the ICU [intensive care unit] and then have a heart or kidney problem…Some countries are recording that as a heart issue or a kidney issue and not a COVID-19 death. The intent is ... if someone dies with COVID-19 we are counting that."[40]

Dr. Scott Jensen, a Minnesota Family practice doctor and state Senator, said that this means that a patient who died after being hit by a bus and tested positive for coronavirus would be listed as having presumed to have died from the virus regardless of whatever damage was caused by the bus.

Dr. Jensen reported receiving a 7-page document from CDC instructing him to do this. As for the motivation? "Fear is a great way to control people," he told a television station.[41]

He was notably outspoken about this matter. He cited situations in the past where he had patients who died while having the flu, stating "I've never been encouraged to [notate 'influenza']. I would probably write 'respiratory arrest' to be the top line, and the underlying cause of this disease would be pneumonia ... I might well put emphysema or congestive heart failure, but I would never put influenza down as the underlying cause of death and yet that's what we are being asked to do here."[42]

When Dr. Anthony Fauci was asked about the number of coronavirus deaths being "padded," he cited the prevalence of "conspiracy theories" during "challenging" times in public health. Dr. Jensen's response to this was "I would remind him that anytime health care intersects with dollars it gets awkward." Dr. Jensen stated that Medicare provides $13,000 to the hospitals and doctors for each COVID-19 patient, much more than the standard for ailments such as influenza, which has averaged around $5,000 in recent years. In addition to that, if a ventilator is used for the patient, Medicare provides $39,000 to the hospital and doctors.[43]

Although Dr. Jensen did not go as far as saying that physicians are trying to pad their pockets, he is more skeptical of those at higher levels such as hospital administrators.

Other misrepresentations about cause of death were being made almost daily. For example, during a press conference, Connecticut Governor Ned Lamont announced that a 6-week-old baby had died and tested positive for coronavirus, and that this was likely one of the youngest deaths from the disease anywhere.[44] His tweet read: "It is with heartbreaking sadness today that we can confirm the first pediatric fatality in Connecticut linked to COVID-19. A 6-week-old newborn from the Hartford area was brought unresponsive to a hospital late last week and could not be revived." He went on to say, "This is a virus that attacks our most fragile without mercy. This also stresses the importance of staying home and limiting exposure to other people. Your life and the lives of others could literally depend on it. Our prayers are with the family at this difficult time."[45]

The problem is that this is not what happened at all. In fact, the state's medical examiner refused to certify death from coronavirus. Toxicology tests are pending, and the medical examiner indicated the possibility that the child had an underlying condition or might have died of sudden infant death syndrome or positional asphyxiation.[46]

But the damage was done. Lamont told the public that "... no one is safe from this virus," and issued this warning, "For those young people who think maybe they're a little more invincible, think again."[47] The public became more frightened, more likely to do as they were told. Stay home, do not congregate, continue to follow directions. He succeeded in scaring people with a false story.

This is not the only example in which a young person was said to have died from COVID-19 when that is not what happened at all. Chloe Middleton, age 21, died from coronavirus, according to her family. She was taken to the hospital after having a heart attack and died shortly after. A coroner said the cause of death was related to COVID-19 because the family reported she had a cough. The hospital had not recorded it as a COVID-19 death because she did not test positive for the disease.

The family took down a Facebook post claiming that Chloe had no underlying health issues and refused to respond to reporters calling for information. Subsequently the coroner's office issued this statement: "Chloe died at Wexham Park Hospital on the 19 March 2020. The case was reported to the Berkshire coroner's office. Her death was very sad but as she had a natural cause of death, involvement by the coroner was not required and the hospital issued a death certificate. There was no postmortem examination or inquest. We must now respect the privacy of her family and cannot provide any further information."[48]

There's More

A study published in April 2020 showed that it is difficult to differentiate between deaths from COVID-19 and Radiation Pneumonitis (RP), which is a common condition that occurs in 15-40% of patients being treated for cancer.[49] Cancer patients are more susceptible to getting the flu and dying from it. We will never know how many were improperly diagnosed or reported, yet it is important to note.

Inaccurate State Death Reports

The *New York Times* reported on April 14, 2020 that New York City had increased its death toll by 3700 people after officials said they would not include people who never tested positive for COVID-19 but were assumed to have it.

After admitting that the cases were not valid, the *Times* reporters wrote, "The numbers brought into clearer focus the staggering toll the virus has already taken on the largest city in the United States, where deserted streets are haunted by the near-constant howl of ambulance sirens."[50]

In Pennsylvania, death rates were adjusted downward when Health Secretary Rachel Levine said on April 23, 2020 that more information is needed before "probable" cases can be attributed to COVID-19. She said the decision was made in the interest of transparency.

This decision resulted in a reduction of 6 deaths in Lehigh Country, and 100 fewer deaths in Philadelphia, 2 fewer in Montgomery County. Bucks county saw a reduction of 10, Monroe county was reduced by 6, and Carbon County was reduced by 2. Total drop was 200 deaths, a significant percentage of the total.[51]

On April 20, 2020, Illinois Department of Health Director Dr. Ngozi Ezike explained how her department decides whether a death is due to COVID-19. She said that anyone who dies and has tested positive is categorized as a COVID-19 death.

Here is, verbatim, what she said:

"If you were in hospice and had already been given a few weeks to live, and then you also were found to have COVID, that would be counted as a COVID death. It means technically even if you died of a clear alternate cause, but you had COVID at the same time, it's still listed as a COVID death. So, everyone who's listed as a COVID death doesn't mean that that was the cause of the death, but they had COVID at the time of the death."[52]

Colorado State representative Mark Baisley has asked for a formal investigation into Jill Ryan, Executive Director of the Colorado Department of Public Health and Environment with the potential for criminal charges to be brought. According to Baisley, Ryan has falsely altered death certificates.

Baisley provided a letter from the Someren Glen senior care facility which was sent to its staff, residents, and families of residents, stating that CPDHE had changed the cause of death recorded by attending physicians in seven cases to reflect COVID-19 instead of the actual cause of death.

The Montezuma County coroner told the same news station that the state overruled the cause of death for a person in his jurisdiction too. The person died of alcohol poisoning, but it was changed to COVID-19.

Eventually the Colorado Department of Health acknowledged that the numbers had been inflated by people who had the virus but died of other causes and adjusted the numbers down from 1150 deaths to 878.[53]

Dr. Deborah Birx, the task force response coordinator, changed her tune about death counts. During a previous White House daily briefing she stated that death certificates were to state COVID-19 as the cause of death if the person tested positive but died of something else. She said the opposite and asked the CDC to exclude from the death count people who had the virus but died of something else and removed those who were presumed to have the virus but did not have confirmed lab results.

Birx and other health officials take issue with the CDC's system now, claiming that the number of cases and mortality may be inflated as much as 25%. "There is nothing from the CDC that I can trust," she told CDC Director Robert Redfield.[54]

In June 2020, Washington State announced a "phased-in" process which would result in telling the truth about COVID-19 deaths. Apparently just telling the truth all at once would be intolerable. The first phase resulted in several suicides, homicides, and overdose deaths being removed from the death count. Health officials also reported that they would categorize deaths as "confirmed, probable, suspect and not COVID."

The Freedom Foundation investigated and reported on this May 18, 2020 after obtaining written data from Washington State DOH officials. When confronted with it, Washington Governor Inslee responded that it was disgusting and malarkey and accused the Freedom Foundation of "fanning these conspiracy claims from the planet Pluto" and not caring about people who died from COVID.[55]

DOH held a press briefing on May 21,2020 during which it confirmed that reported deaths were inflated and that "(w)e currently do have some deaths that are being reported that are clearly from other causes" including some "...from gunshot wounds."

Some "Deaths" Were Clearly NOT COVID!

Coal miner Nathan Turner was 30 years old when he was found dead in his home by his fiancé in Queensland, Australia. Queensland Health promptly reported that Turner died of coronavirus and claimed that he was Australia's youngest COVID-19 victim. Local doctors reported that Turner's death baffled them as he had not left his small town since February. They hypothesize that perhaps a nurse from 400 km away who had driven to Blackwater to watch the sunset had infected him.

After all of this, autopsy showed that Turner did not have the virus. The family was furious and called on Premier Annastacia Palaszczuk and health official Jeannette Young to apologize to both the family and to the community for creating "chaos and panic."

"You should be ashamed of yourself and if you had any human decency left then you will apologise for creating trauma to this family whilst you caused panic to our community.

"This is unacceptable behaviour from our leaders in power who forced a family to sit in silence and not to comment about the chaos they were about to inflict on our state."

Queensland Health admits administering additional tests which also were negative for COVID. Apparently, there are many who are intent on making a diagnosis of COVID even when it is not there.

An online petition demanding a truthful apology had gathered 2092 signatures out of a 2500 goal within just a few hours.[56]

One of the more insane episodes of deaths categorized as COVID-19 involved a man who was shot by the NYPD after threatening officers with a knife and gun.

Ricardo Cardona called 911 on himself and then repeatedly told officers to kill him when they arrived to find him with the weapons. He later told investigators that he wanted to die by suicide by cop since he had been infected with COVID-19. The officers ultimately fired 11 shots, 7 of which hit him. He died 5 days later, and his death is attributed to COVID-19 with his wounds and underlying health conditions listed as "complicating factors."[57]

ENDNOTES

1. David Pride. Hundreds of different coronavirus tests are being used – which is best? *The Conversation* April 4 2020 https://www.marketwatch.com/story/hundreds-of-different-coronavirus-tests-are-being-used-which-is-best-2020-04-02 accessed 9.2.2020

2. IBID

3. Celia Farber. Was the COVID-19 Test Meant to Detect a Virus?" April 7 2020 https://uncoverdc.com/2020/04/07/was-the-covid-19-test-meant-to-detect-a-virus/ accessed 7.2.2020

4. Gina Kolata. Faith in Quick Test Leads to Epidemic That Wasn't. *New York Times* Jan 22 2007 https://www.nytimes.com/2007/01/22/health/22whoop.html accessed 9.2.2020

5. Watson J, Whiting PF, Brush JE. "Interpreting a covid-19 test result." *BMJ* 2020 May;369:m1808

6. Engelbrecht T, Demeter K. "COVID19 PCR Tests are Scientifically Meaningless." Bulgarian Pathology Association. Jan 7 2020 https://bpa-pathology.com/covid19-pcr-tests-are-scientifically-meaningless/ accessed 9.2.2020

7. Fermin Koop. A startling number of coronavirus patients get reinfected. *ZME Science* Feb 26 2020 https://www.zmescience.com/science/a-startling-number-of-coronavirus-patients-get-reinfected/ accessed 9.2.2020

8. Li Y, Yao L, Li J et al. "Stability issues of RT-PCR testing of SARS-CoV-2 for hospitalized patients clinically diagnosed with COVID-19." *J Med Virol* 2020 Jul;92(7):903-908

9. Coco Feng, Minghe Hu. Race to diagnose coronavirus patients constrained by shortage of reliable detection kits. *South China Morning Post* Feb 11 2020 https://www.scmp.com/tech/science-research/article/3049858/

race-diagnose-treat-coronavirus-patients-constrained-shortage accessed 9.2.2020

10. CDC 2019-Novel Coronavirus (2019-nCoV) Real-Time RT-PCR Diagnostic Panel. Centers for Disease Control and Preention. https://www.fda.gov/media/134922/download accessed 9.2.2020

11. ACCELERATED EMERGENCY USE AUTHORIZATION (EUA) SUMMARY COVID-19 RT-PCR TEST (LABORATORY CORPORATION OF AMERICA) U.S. Food and Drug Administration. https://www.fda.gov/media/136151/download accessed 9.2.2020

12. BIO-RAD SARS-CoV-2/Covid-19 Diagnosis and Confirmation Solutions. https://www.bio-rad.com/featured/en/sars-cov-2-covid-19-testing-solutions.html accessed 9.2.2020

13. Michael Mendizza. Why The Coronavirus Will Soon Vanish Overnight. https://ttfuture.org/blog/michael/why-coronavirus-will-soon-vanish-overnight accessed 9.2.2020

14. Ana Radelat. Lamont, other governors, seek $500 billion in new coronavirus stimulus money for states. *The CT Mirror* https://ctmirror.org/2020/04/16/lamont-other-governors-seek-500-billion-in-new-coronavirus-stimulus-money-for-states/ accessed 9.2.2020

15. Monthly Federal Soending/Revenue/Deficit Charts Federal Coronavirus/COVID-19 Response. https://www.usgovernmentspending.com/compare_state_debt 9.2.2020

16. Has COVID-19 Testing Made the Problem Worse? Confusion Regarding "The True Health Impacts". Centre for Research on Globalization. https://www.globalresearch.ca/has-covid-19-testing-made-the-problem-worse-confusion-regarding-the-true-health-impacts/5709323 accessed 9.2.2020

17. Beth Mole. CDC's failed coronavirus tests were tainted with coronavirus, feds confirm. *Ars Technica* April 20 2020 https://arstechnica.com/science/2020/04/

cdcs-failed-coronavirus-tests-were-tainted-with-coronavirus-feds-confirm/ accessed 9.2.2020

18. Jorge Gonzalez-Gallarza Hernandez. China challenges the world with flawed COVID-19 test kits. March 30 2020. https://www.washingtontimes.com/news/2020/mar/30/china-challenges-the-world-with-flawed-covid-19-te/ accessed 9.2.2020

19. Laura Terrell. 'False negatives are harmful' according to medical professionals. *KCCI* April 3 2020 https://www.kcci.com/article/false-negatives-are-harmful-according-to-medical-professionals/32038917 accessed 9.2.2020

20. Amanda Morris. People look to COVID-19 antibody testing for answers, but no test offers guarantees. *Azcentral* April 27 2020 https://www.azcentral.com/story/news/local/arizona-health/2020/04/27/questions-linger-covid-19-antibody-tests-even-demand-grows/5170052002/ accessed 9.2.2020

21. Associated Press. FDA tightens rules on antibody test after false claims, accuracy problems. *NBC News* May 4 2020 https://www.nbcnews.com/health/health-news/fda-tightens-rules-antibody-tests-after-false-claims-accuracy-problems-n1199431 accessed 9.2.2020

22. Alexis Madrigal, Robinson Meyer. 'How Could the CDC Make That Mistake?' *The Atlantic* May 21 2020 https://www.theatlantic.com/health/archive/2020/05/cdc-and-states-are-misreporting-covid-19-test-data-pennsylvania-georgia-texas/611935/ accessed 9.2.2020

23. Ben Cost "Faulty Coronavirus Kits suspected as goat and fruit test positive in Tanzania" *New York Post* May 6 2020 https://nypost.com/2020/05/06/faulty-coronavirus-kits-suspected-as-goat-and-fruit-test-positive-in-tanzania/ accessed 9.2.2020

24. Tanzania COVID-19 lab head suspended as president questions data. Al Jazeera May 5 2020 https://www.aljazeera.com/news/2020/05/tanzania-covid-19-lab-head-suspended-president-questions-data-200505065136872.html accessed 9.2.2020

25. Maura Hohman.. NBC's Dr. Joseph Fair hospitalized with corona-virus: 'Not out of the woods yet.' *Today* May 13 2020 https://www.today.com/health/nbc-news-contributor-dr-joseph-fair-sick-coronavi-rus-t181487 accessed 9.2.2020

26. Heather Boerner. COVID-19 Test Results: Don't Discount Medical Intuition. *Medscape* May 16 2020 https://www.medscape.com/view-article/930650 accessed 9.2.2020

27. FDA News Release. Coronavirus (COVID-19) Update: FDA Informs Public About Possible Accuracy Concerns with Abbott ID NOW Point-of-Care Test. U.S. Food and Drug Administration. May 14 2020 https://www.fda.gov/news-events/press-announcements/coronavirus-covid-19-update-fda-informs-public-about-possible-ac-curacy-concerns-abbott-id-now-point

28. Kucirka LM, Lauer SA, Laeyendecker O, Boon D, Lessler J. "Variation in False-Negative Rate of Reverse Transcript Polymerase Chain Reaction –Based SARS-CoV-2 Tests by Time of Exposure." *Ann Intern Med* 2020 May;M20-1495

29. Heather Boerner. COVID-19 Test Results: Don't Discount Medical Intuition. *Medscape* May 16 2020 https://www.medscape.com/view-article/930650 accessed 9.2.2020

30. Arevalo-Rodriguez I, Buitrago-Garcia D, Simancas-Racines D et al. "FALSE NEGATIVE RESULTS OF INITIAL RT-PCR ASSAYS FOR COVID-19: A SYSTEMATIC REVIEW." *MedRxiv* doi: https://doi.org/10.1101/2020.04.16.20066787

31. Mason Boycott-Owen, Paul Nuki. Tens of thousands of coro-navirus tests have been double-counted, officials admit. *The Telegraph* May 21 2020 https://www.telegraph.co.uk/global-health/science-and-disease/tens-thousands-coronavirus-tests-have-dou-ble-counted-officials/ accessed 9.2.2020

32. Coronavirus Disease 2019 (COVID-19). 2020 Interim Case Definition, Approved April 5 2020. Canters for Disease Control and

Prevention. https://wwwn.cdc.gov/nndss/conditions/coronavirus-dis-ease-2019-covid-19/case-definition/2020/ accessed 9.2.2020

33. Menni C, Sudre CH, Steves CJ, Ourselin S, Spector TD. "Quantifying additional COVID-19 symptoms will save lives." *Lancet* published online June 4 2020

34. Weige-Lussen A, Wolfensberger M. "Olfactory Disorders following Upper Repirtory Tract Infections." In Hummel T, Welge-Lüssen A (eds): Taste and Smell. An Update. Adv Otorhinolaryngol. Basel, Karger, 2006, vol 63, pp 125-132

35. Boesveldt S, Postma EM, Boak D et al. "Anosmia – A Clinical Review." *Chem Senses* 2017 Sep;42(7):513-523.

36. Menni C, Sudre CH, Steves CJ, Ourselin S, Spector TD. "Quantifying additional COVID-19 symptoms will save lives." *Lancet* 2020 Jun;395(10241):E107-E108

37. Sarah Elizabeth Richards. "Lost your sense of smell? It may not be coronavirus." *National Geographic* April 7 2020 https://www.nationalgeographic.com/science/2020/04/lost-your-sense-of-smell-it-may-not-be-coronavirus/ accessed 9.2.2020

38. Guidance for Certifying Deaths Due to Coronavirus Disease 2019 (COVID-19). Vital Statistics Reporting Guidance. Report no. 3 April 2020 https://www.cdc.gov/nchs/data/nvss/vsrg/vsrg03-508.pdf accessed 9.2.2020

39. New ICD code introduced for COVID-19 deaths. COVID-19 Alert No. 2 March 24 2020. National Vital Statistics System. https://d33wjekvz3zs1a.cloudfront.net/wp-content/uploads/2020/04/Alert-2-New-ICD-code-introduced-for-COVID-19-deaths.pdf accessed 9.2.2020

40. Louis Casiano. Birx says government is classifying all deaths of patients with coronavirus as 'COVID-19' deaths, regardless of cause. *Fox News* April 7 2020 https://www.foxnews.com/politics/

birx-says-government-is-classifying-all-deaths-of-patients-with-coronavirus-as-covid-19-deaths-regardless-of-cause accessed 9.2.2020

41. MN Sen and Dr. Scott Jensen said that he received a 7 pg doc from MN Health to fil out death… *FOX News* April 8 2020 https://www.youtube.com/watch?v=Pfa4b7T0ZHY accessed 9.2.2020

42. Charles Creitz. Minnesota doctor blasts 'ridiculous' CDC coronavirus death count guidelines. *Fox News* April 9 2020 https://www.foxnews.com/media/physician-blasts-cdc-coronavirus-death-count-guidelines accessed 9.2.2020

43. Michelle Rogers. Fact check: Hospitals get paid more if patients listed as COVID-19, on ventilators. *USA Today* April 24 2020 https://www.usatoday.com/story/news/factcheck/2020/04/24/fact-check-medicare-hospitals-paid-more-covid-19-patients-coronavirus/3000638001/ accessed 9.2.2020

44. Audrey McNamara. 6-week-old baby's death linked to coronavirus, believed to be one of the youngest fatalities. *CBS News* April 2 2020 https://www.cbsnews.com/news/six-week-old-baby-dies-coronavirus-believed-to-be-youngest-fatality/ accessed 9.2.2020

45. Ariel Zilber. Coroner refuses to rule COVID-19 as cause of death of six-week-old baby after Connecticut governor claimed toddler was 'youngest coronavirus victim in the world. *Daily Mail* April 6 2020 https://www.dailymail.co.uk/news/article-8193487/Coroner-refuses-rule-COVID-19-cause-death-six-week-old-Connecticut-baby.html accessed 9.2.2020

46. IBID

47. IBID

48. Matthew Weaver. Chloe Middleton; death of 21-year-old not recorded as Covid-19. *The Guardian* March 27 2020 https://www.inkl.com/news/chloe-middleton-death-of-21-year-old-not-recorded-by-nhs-as-covid-19-related accessed 9.2.2020

49. Shaverdian N, Shepherd AF, Rimner A et al "Need for Caution in the Diagnosis of Radiation Pneumonitis During the COVID-19 Pandemic. *Adv Radiat Oncol* 2020 Jul-Aug;5(4):617-620

50. J. David Goodman and William K. Rashbaum. N.Y.C. Death Toll Soars Past 10,000 in Revised Virus Count. *New York Times* April 21 2020 https://www.nytimes.com/2020/04/14/nyregion/new-york-coronavirus-deaths.html accessed 9.2.2020

51. Steve Novak. Pa. coronavirus update: New cases, rise, but death toll drops? Here's why, and what it means in the LeHigh Valley. *lehighvalleylive.com* April 23 2020 https://www.lehighvalleylive.com/coronavirus/2020/04/pa-coronavirus-update-new-cases-rise-but-death-toll-drops-heres-why-and-what-it-means-in-the-lehigh-valley-covid-19-case-map-42320.html accessed 9.2.2020

52. IDPH Director explains how Covid deaths are classified. *25News Week.com* April 20 2020 https://week.com/2020/04/20/idph-director-explains-how-covid-deaths-are-classified/ accessed 9.2.2020

53. Kyle Clark. GOP rep alleges falsified COVID-19 records, calls for indictment of Colorado's top health official. *9News* May 14 2020 https://www.9news.com/article/news/local/next/gop-rep-alleges-falsified-covid-19-records-calls-for-indictment-of-colorados-top-health-official/73-bf02452f-4615-4efe-9413-a4826a8105b2 accessed 9.2.2020

54. Joseph Guzman. "Trump administration pushing CDC to change how it counts coronavirus deaths: report." *The Hill* https://thehill.com/changing-america/well-being/longevity/497602-trump-administration-pushing-cdc-to-change-how-it accessed 9.2.2020

55. Maxford Nelsen. WA Dept. of Health to stop counting deaths improperly attributed to COVID-19. *Freedom Foundation* Jun 17 2020 https://www.freedomfoundation.com/washington/wa-dept-of-health-to-stop-counting-deaths-improperly-attributed-to-covid-19/ accessed 9.2.2020

56. Alana Mazzoni. Furious family of Nathan Turner, 30, demand an apology after authorities declared the miner was 'Australia's youngest coronavirus victim' - but tests reveal he DIDN'T HAVE the virus. *Daily Mail Australia* June 1 2020 https://www.dailymail.co.uk/news/article-8376959/Nathan-Turners-family-demand-apology-declared-Australias-youngest-coronavirus-victim.html accessed 9.2.2020

57. Michel R. Sisak. "NYPD: Man shot by officers later dies of coronavirus." *ABC News* May 20 2020 https://abcnews.go.com/US/wireStory/nypd-man-shot-officers-dies-coronavirus-70941694 accessed 9.2.2020

THE CONTROVERSY OVER TREATMENT

TREATMENT OPTIONS for patients with COVID-19 created just as much hysteria as the virus itself, and much of the controversy concerned chloroquine and hydroxychloroquine (a less toxic derivative of chloroquine), two inexpensive drugs used to treat malaria for decades. In March President Trump announced that hydroxychloroquine (HCQ) could be a potential game changer in the fight against COVID-19. Hysteria ensued. The FDA warned that the drug could cause serious heart problems for COVID-19 patients.[1]

Fauci also disagreed with Trump, stating that Americans should be cautious about hydroxychloroquine, and that definitive studies were needed to show that it was effective for flu viruses. At that time, he stated that the "…only effective tool we have for fighting the coronavirus is social distancing," and urged Americans to stay indoors.[2]

It's hard to imagine that Fauci, touted as one of the world's best infectious disease experts, did not know that chloroquine had been found to be effective for treating SARS infections and preventing spread in 2005,[3] and that the CDC posted a study showing it to be effective on its website.[4] Additionally, in vitro studies dating back to 2004 showed that chloroquine was effective for inhibiting virus spread.[5] Research on the use of HCQ for the treatment of COVID-19 started almost immediately when the virus was declared a pandemic, and much of this research was positive.

Yet in a Town Hall on CNN, Fauci stated "There are no proven safe and effective therapies for the coronavirus."[6]

Research Findings

More recent studies starting in March 2020 showed that an HCQ/antibiotic combination was effective for treatment of COVID-19.[7]

A retrospective study conducted in China reported that low-dose hydroxychloroquine "was associated with the reduced fatality of critically ill patients with COVID-19." Laboratory data also showed that the medication "greatly lowered the levels of IL-6, one of the most inflammatory cytokines." The study included 550 critically ill COVID-19 patients who required mechanical ventilation. The dose used was 200 mg two times daily for 7-10 days, a dose which has been safely used for the treatment of malaria for decades. The researchers wrote, "HCQ treatment significantly reduced the fatality of critically ill COVID-19 patients" as compared with patients who received only basic treatments, and that the drug worked "…without apparent toxicity." Additionally, the researchers recommended that HCQ be used as "an option for a patient at an early stage considering its safety records and the long history of its use in treating malaria infections."[8]

A systematic review conducted in March 2020 summarized available evidence for the use of HCQ for COVID-19 and concluded, "There is theoretical, experimental, preclinical and clinical evidence of the effectiveness of chloroquine in patients affected with COVID-19. There is adequate evidence of drug safety from the long-time clinical use of chloroquine and hydroxychloroquine in other indications."[9]

In multicenter clinical trials in China, chloroquine was showed to be safe and effective against COVID-19 associated pneumonia. The researchers wrote, "The drug is recommended to be included in the next version of the Guidelines for the Prevention, Diagnosis, and Treatment of Pneumonia Caused by COVID-19 issued by the National Health Commission of the People's Republic of China for treatment of COVID-19 infection in larger populations in the future."[10]

A multicenter prospective observational study of COVID-19 patients concluded that the median time to achieve undetectable viral RNA was shorter in the chloroquine group than in the controls treated with other drugs.[11]

An analysis of the medical records of 166 COVID-19 patient showed that 48.8% of patients not treated with HCQ died, while only 22% of those treated with HCQ died.[12]

Doctors Reported Success

Dr. Vladimir Zelenko reported using a combination of HCQ, zinc and azithromycin on hundreds of patients with a 99.9% success rate. Only one outpatient died, and this person did not follow the protocol.[13]

Dr. Simone Gold found HCQ to be fast-acting, which she found surprising since infectious diseases do not usually respond to treatment quickly. She prescribed it to a few patients, and they got better in 12 hours. She became an advocate for HCQ but received a lot of pushback from the hospital, including advice to stop prescribing it to outpatients. She stated that there are no safety issues – the drug has been used safely for 65 years and reported that the response of the hospital and her colleagues was "baffling."[14]

Dr. Anthony Cardillo, CEO of Mend Urgent Care, told KABC-TV, "Every patient I've prescribed it to has been very, very ill and within 8 to 12 hours, they were basically symptom-free. So, clinically I am seeing a resolution." He added that combining the drug with zinc has been the key to the success. The hydroxychloroquine, he said, "opens the zinc channel" allowing the zinc to enter the cell, which then "blocks the replication of cellular machinery."[15]

Dr. Ivette Lozano in Dallas Texas started prescribing a combination of HCQ, zinc and azithromycin to her patients with excellent results. Within only 6-8 hours most were better. But on March 20, 2020, the Texas State Board of Pharmacy issued a rule that no prescriptions for HCQ or azithromycin could be issued without a diagnosis "consistent with evidence for its use." According to Lozano, "Never before have we had to turn in a diagnosis with a prescription."

Lozano contacted State Senator Bob Hall concerning this rule, with her concerns about patient privacy, and the fact that the rule seemed to discourage the use of a potentially life-saving drug. Hall agreed and added his concern about "collusion be-tween the pharmacy board and pharmaceutical companies who want to prevent the use of an inexpensive drug while they de-velop a new, expensive drug." After he contacted the pharmacy board, the rule was rolled back.[16]

The Texas pharmacy board was not the only state to issue such as restriction. Most others issued some sort of restrictive and even threatening guidelines concerning the prescribing of HCQ. Several boards specifically forbade the prescribing of HCQ for prophylaxis. Many required that pharmacists demand that prescriptions be accompanied by a diagnosis, and specifical-ly stated that the drug should not be prescribed off-label. Several reminded pharmacists that there were no FDA-approved drugs to treat COVID-19 and a few included veiled threats to doctors. California, for example, issued a warning to "remind health care professionals that inappropriately prescribing or dispensing medications constitutes unprofessional conduct in California. Prescribers and pharmacists are obligated to follow the law, stan-dard of care, and professional codes of ethics in serving their patients and public health."[17]

In short, it seemed there was an outright campaign to dis-credit the use of HCQ for COVID-19 patients.

Digging in Their Heels

On Monday, May 17, 2020, President Trump announced he was taking hydroxychloroquine, with the approval of the White House Physician. He said, "All I can tell you is so far I seem to be OK," adding that he had been taking the drug for about a week and a half, with the approval of the White House physician. "I get a lot of tremendously positive news on the hydroxy," Mr. Trump continued, explaining that his decision to try the drug was based on one of his favorite refrains: "What do you have to lose?"[18]

The response from the medical community was hysteria. Dr. Manny Alvarez, senior managing editor of Fox News' Health News department said in a tweet that Trump was irresponsible when he told the public he was taking the drug and demanded that the White House Physician explain to the public what changed "...since studies showed no benefit."[19] Dr. Scott Solomon, professor of Medicine at Harvard Medical School, also said that Trump was irresponsible for announcing that he was taking the drug. And Steve Nissen at the Cleveland Clinic warned that there were "serious hazards" associated with taking hydroxychloroquine.[20]

Fauci dug in his heels, insisting that HCQ was not an effective treatment for COVID-19. Fauci told CNN, "The scientific data is really quite evident now about the lack of efficacy for it," and added that there was a risk of "adverse events with regard to cardiovascular."[21]

Anything to Discredit the Cheaper Cure

On May 22, 2020 the *Lancet* published a study reporting that hydroxychloroquine and chloroquine (the older version of the drug) were linked to increased deaths in hospitals all over the world.[22] The paper was authored by Dr. Mandeep R. Mehra of Harvard Medical School, and Dr. Sapan S. Desai of Surgisphere, an Illinois-based company that had supposedly gathered data on tens of thousands of patients in over 1000 hospitals worldwide.

The authors reported that their data represented over 15,000 patients who received hydroxychloroquine or chloroquine and over 81,000 patients who did not receive these drugs. According to this paper, one in six patients taking only one of these drugs died; one in five taking chloroquine with an antibiotic died; and one in four taking hydroxychloroquine with an antibiotic died. The death rate for patients not taking these drugs was one in eleven. Additionally, serious arrhythmias were reported, with most occurring in the group taking hydroxychloroquine in combination with an antibiotic (8% versus 0.3% in patients not given any of these drugs or combinations).

The mainstream media was thrilled, and this became a big story. President Trump had made positive comments about hydroxychloroquine during at least one of the daily White House press briefings and had also reported taking it himself. This was evidence that the president had misled the country when he talked about HCQ.

Fauci had repeatedly made negative remarks about the drug. Even though Fauci has been wrong about practically everything since the COVID-19 nonsense began, he was still adored by members of the media, who seemed to wait with bated breath for each new erroneous statement he uttered. The *Lancet* study fit in well with the false but repeated narrative – Trump is always wrong and Fauci is always right, and HCQ was a bad choice for treating COVID-19.

On May 27, 2020, just days after the *Lancet* study was published, Fauci made this statement on CNN, "The scientific data is really quite evident now about the lack of efficacy."[23]

Dr. Birx chimed in "…it clearly shows that co-morbidity that puts individuals at more risk. And I think it's one of our clearest studies because there were so many, tens of thousands of individuals involved; that the doctors clearly annotated who had heart disease and who had obesity. And you could see dramatically the increased risk…"[24]

The *Lancet* paper had a major impact on both policy and research. The WHO stopped the hydroxychloroquine arm of its clinical trials.[25] Research studies using hydroxychloroquine in the UK and France were also halted. The COVID-19 storytellers repeatedly reported that the issue was settled. Science showed that hydroxychloroquine was a dangerous drug, and not appropriate for the treatment of COVID-19.

The Guardian was one of the first media outlets to question the *Lancet* study. According to the article, data from five Australian hospitals with 600 COVID-19 patients and 73 deaths were included in the analysis. But at the time the data was collected, there had been only 67 deaths recorded throughout Australia. *The Guardian* was able to confirm that the National Notifiable Diseases Surveillance System was not the source of the information. Health departments in New South Wales and Victoria, two of Australia's most populous states, stated that the reported data did not reconcile with their data and that they did not provide any data to the researchers who claimed to have gathered data.[26]

Guardian investigative reporters also looked into Surgisphere and reported that one of the firm's science editors appeared to be a science fiction writer and fantasy artist. One of the company's marketing executives also had a career as an adult model and events hostess. The company's LinkedIn page shows only three employees as of June 3, 2020, making it highly unlikely that the company had the resources to gather and analyze such a large data base, which consisted of 96,032 patients who were admitted to hundreds of hospitals on six continents by April 14, 2020. Additionally, *Guardian* reports that until June 1, 2020, the "get in touch" link on Surgisphere's website led to a cryptocurrency website.[27]

Researchers and writers at *The Scientist* also reported concerns about the study. The *Lancet* article reported that Surgisphere's registry included data from over 63,000 COVID patients admitted to 559 hospitals in North America by April 14, 2020. But Surgisphere CEO and founder Sapan Desai refused to provide the names of any of the hospitals when asked. *The Scientist* contacted some of the larger health systems in states reporting the most cases and deaths and did not find any who confirmed that they provided data to Surgisphere.[28]

Other researchers interviewed by *The Scientist* had doubts about the African data, noting that the quality of electronic health records in Africa made it highly unlikely that records for 4402 hospitalized patients could have been obtained from African countries by April 14, when at the time only 15,738 cases had been reported on the entire continent.[29]

At the same time, another article published in the *New England Journal of Medicine* reported that patients with COVID-19 and with cardiovascular disease had an increased risk of dying in the hospital. This article reported data from 346 COVID-19 patients hospitalized in Turkey by March 15.[30]

Open letters signed by over 140 scientists and physicians were sent to both the *Lancet* and the *NEJM*. The letter to the *NEJM* states that "countrywide, the first COVID-19 case was diagnosed at Istanbul Faculty of Medicine on the 9th of March. The second COVID-19 patient in that hospital was not seen until the 16th of March. The Turkish Ministry of Health reported a total of only 191 PCR positive cases by the 18th of March."[31] In other words, the *NEJM* article reported more COVID-19 patients in Turkey than had been diagnosed at the time.

The letter to the *Lancet* expressed "both methodological and data integrity concerns" and listed, among other issues:

1. The study's authors did not indicate the "severity" of the disease being treated. Was it early on in the COVID-19 progression or late in the process? The dosages of HCQ or CQ used were not disclosed.
2. The authors have not adhered to "standard practices in the machine learning and statistics community. They have not released their code or data. There is no data/code sharing and availability statement in the paper."
3. The countries and hospitals from which the data were obtained were not disclosed, and the authors have denied requests for that information.

4. The numbers of cases and deaths as well as the detailed data collection from Surgisphere-associated hospitals in Africa "seem unlikely."

5. Reported ratios of HCQ to CQ are "implausible."[32]

Both journals expressed concern about the data they published. On June 4, 2020, three of the authors of the *Lancet* paper retracted their study, claiming they were "unable to complete an independent audit of the data underpinning their analysis"…and "…that they can no longer vouch for the veracity of the primary data sources."[33] The *NEJM* article was also retracted.[34] The World Health Organization has resumed its research on the use of hydroxychloroquine for the treatment of COVID-19.

The media reminded us many times every day that Fauci was the world's best virologist, and featured clips of his mantra about always following the science. It's difficult to believe that such a great doctor and scientist could be taken in by such an overtly fraudulent article. Seemingly the reporters at *the Guardian* were better at reviewing the science than Fauci.

Fauci's Favorite Treatment

Remdesivir is a non-specific antiviral drug with both safety and efficacy concerns, which costs thousands of dollars per treatment course. A trial of four different treatments for Ebola was conducted during an outbreak in The Democratic Republic of Congo in 2018. The data and safety board recommended that patients taking remdesivir be assigned to one of the other drugs due to increased mortality rates in the remdesivir group.[35]

Despite this, The National Institutes of Health started a randomized controlled trial with 1063 COVID-19 patients taking either remdesivir or placebo on February 21, 2020.

A study published in the *Lancet* showed that remdesivir did not reduce death rates or even make COVID-19 patients feel better. Furthermore, 12% of the patients in the remdesivir group dropped out due to serious side effects such as acute respiratory distress syndrome or respiratory failure.[36]

Fauci claimed that the results of the NIH trial were so promising that there was "an ethical obligation to immediately let the placebo group know so they can have access" to the drug.[37] As a result the trial was unblinded and the actual results will never be known. While this is only speculation, this may have been done to avoid the same disastrous outcomes reported in the *Lancet* trial.

Indeed, it appears that Fauci had already made up his mind about remdesivir, and even changed the primary metric for measuring outcomes two weeks before he announced that the drug would be the "new standard of care." Instead of reporting how many people taking remdesivir were kept alive on ventilators or died, Fauci's agency would evaluate the drug based on how long it took patients to recover.[38] This was announced on the clinical-trials.gov website, but received almost no attention at the time.

Of course, it gets worse. Even the definition of "recovery" was manipulated. Time to recovery was defined as "the first day on which the subject satisfies one of the following three categories from the ordinal scale: 1) Hospitalized, not requiring supplemental oxygen – no longer requires ongoing medical care; 2) Not hospitalized, limitation on activities and/or requiring home oxygen; 3) Not hospitalized, no limitations on activities."[39] This means that a person who remains hospitalized or who returns home and still requires oxygen treatment is reported as a "success."

Why would Fauci do this? Fauci assembled a 50-member panel to investigate treatment guidelines for COVID-19. Not surprisingly, several members of the panel had ties to pharmaceutical companies. The company with the most members, a total of nine, was Gilead, the manufacturer of remdesivir.[40] This was a sweetheart deal for Gilead. Fauci's agency, the NIAID, organized and paid for the remdesivir trial, saving the company hundreds of thousands of dollars.

Not surprisingly, the panel warned against the use of HCQ and azithromycin for patients outside clinical trials, stating that the drug combo might do more harm than good.[41]

The decision to make sure remdesivir would be the "best choice" seems to have been determined by Fauci before the start of the trial, as is evidenced by both his dismissal of HCQ and the change in primary endpoints in the remdesivir study, a practice considered unethical by many researchers.[42] Establishing endpoints in advance is a fundamental requirement of independent research. Failure to do so, or changing endpoints after a research study begins, indicates potential bias and/or manipulation of study results. This seems to be what happened here.

Unfortunately, scientific misconduct is not unusual. What makes this episode more disgusting is that it was engineered by a doctor who proclaimed publicly that "I give advice according to the best scientific evidence" when testifying in front of a Senate panel.[43]

The best scientific evidence seems to support the use of HCQ in combination with an antibiotic and perhaps zinc. Thousands of doctors all over the world reported incredible success treating patients with these inexpensive combinations. But if an inexpensive cure exists, there would be no need for a vaccine, and certainly no need to keep schools and businesses closed. Perhaps that is really the issue here.

It is unfortunate that Fauci is still a darling of the media, held up as an example of virtue, scientific integrity, and strength. As it turns out, the President, with no medical training at all, seems to have a better handle on the best treatment for COVID-19 patients.

THE CONTROVERSY OVER TREATMENT

ENDNOTES

1. FDA Drug Safety Communication. U.S. Food and Drug Administration. https://www.fda.gov/media/137250/download accessed 6.13.2020

2. Berkley Lovelace Jr. Coronavirus: Dr. Anthony Fauci warns that Americans shouldn't assume hydroxychloroquine is a 'knockout drug.' *CNBC* April 3 2020 https://www.cnbc.com/2020/04/03/coronavirus-fauci-warns-americans-shouldnt-assume-hydroxychloroquine-is-a-knockout-drug.html accessed 9.2.2020

3. Vincent MJ, Bergeron E, Benjannet S et al. "Chloroquine is a potent inhibitor of SARS coronavirus infection and spread." *Virology J* 2005 Aug;69:2:69

4. Chloroquine is a potent inhibitor of SARS coronavirus infection and spread. CDC Stacks Public Health Publications. Centers for Disease Control and Prevention. https://stacks.cdc.gov/view/cdc/3620 accessed 9.2.2020

5. Keyaerts E, Vijgen L, Maes P, Neyts J, Van Ranst M. "In Vitro Inhibition of Severe Acute Respiratory Syndrome Coronavirus by Chloroquine." *Biochem Biophys Res Commun* 2004 Oct 8;323(1):264-8

6. Amanda Woods. Dr. Anthony Fauci says 'there's no magic drug' to treat coronavirus. *New York Post* March 20 2020 https://nypost.com/2020/03/20/dr-anthony-fauci-says-theres-no-magic-drug-to-treat-coronavirus/ accessed 9.2.2020

7. Gautret P, Lagier JC, Parola P et al. "Hydroxychloroquine and azithromycin as a treatment of COVID-19: results of an open-label non-randomized clinical trial." *Int J Antimicrob Agents* 2020 Jul;56(1):105949

8. Yu B, Li C, Chen P. "Low dose of hydroxychloroquine reduces fatality of critically ill patients with COVID-19." *Sci China Life Sci* 2020 May 15;1-7 doi: 10.1007/s11427-020-1732-2. Online ahead of print.

9. Kapoor KM, Kapoor A. "Role of Chloroquine and Hydroxychloroquine in the Treatment of COVID-19 Infection- A Systematic Literature Review." medRxiv doi: https://doi.org/10.1101/2020.03.24.20042366

10. Gao J, Tian Z, Yang X. "Breakthrough: Chloroquine phosphate has shown apparent efficacy in treatment of COVID-19 associated pneumonia in clinical studies." *BioSci Trends* 2020 Mar;14(1):72-73

11. Huang M, Li M, Xiao F et al. "Preliminary evidence from a multicenter prospective observational study of the safety and efficacy of chloroquine for the treatment of COVID-19." *Nat Sci Rev* 2020 May; nwaa113, https://doi.org/10.1093/nsr/nwaa113

12. De Novales FJM, Ramirez-Olivencia G, Estebanez M et al. "Early Hydroxychloroquine Is Associated with an Increase of Survival in COVID-19 Patients: An Observational Study." *Preprints* 2020 May; doi: 10.20944/preprints202005.0057.v1

13. Nickie Louise. New updates from Dr. Vladimir Zelenko: Cocktail of Hydroxychloroquine, Zinc Sulfate and Azithromycin are showing phenomenon results with 900 coronavirus patients treated – Must Watch Video. Tech Startups April 5 2020 https://techstartups.com/2020/04/05/new-updates-dr-vladimir-zelenko-cocktail-hydroxychloroquine-zinc-sulfate-azithromycin-showing-phenomenon-results-900-coronavirus-patients-treated-must-watch-video/ accessed 6.14.2020

14. A Malaria Drug For COVID-19 - Yes, It Works, Says A Doctor. Dennis Prager Show April 14 2020 https://am870theanswer.com/podcast/episode/410/a-malaria-drug-for-covid-19-yes-it-works-says-a-do accessed 9.2.2020

15. Louise N. "More doctors are seeing success with hydroxychloroquine and Zinc Sulphate in treating coronavirus patients." April 6 2020 https://techstartups.com/2020/04/06/more-doctors-are-seeing-success-with-hydroxychloroquine-and-zinc-sulphate-in-treating-coronavirus-patients/ accessed 6.14.2020

16. Kim Roberts. Pharmacy Board Loosens Restrictions on Hydroxychloroquine Prescriptions, Reversing Course. *The Texan* May 15 2020 https://thetexan.news/pharmacy-board-loosens-restrictions-on-hydroxychloroquine-prescriptions-reversing-course/ accessed 9.2.2020

17. COVID-19: Information from the States. https://naspa.us/resource/covid-19-information-from-the-states/ accessed 9.2.2020

18. Annie Karni and Katie Thomas. Trump Says He's Taking Hydroxychloroquine, Prompting Warning From Health Experts. *New York Times* May 18 2020 https://www.nytimes.com/2020/05/18/us/politics/trump-hydroxychloroquine-covid-coronavirus.html accessed 9.2.2020

19. https://twitter.com/justinbaragona/status/1262508060410613765 accessed 9.2.2020

20. Annie Karni and Katie Thomas. Trump Says He's Taking Hydroxychloroquine, Prompting Warning From Health Experts. *New York Times* May 18 2020 https://www.nytimes.com/2020/05/18/us/politics/trump-hydroxychloroquine-covid-coronavirus.html accessed 9.2.2020

21. Devan Cole. Fauci: Science shows hydroxychloroquine is not effective as a coronavirus treatment. *CNN* May 27 2020 https://www.cnn.com/2020/05/27/politics/anthony-fauci-hydroxychloroquine-trump-cnntv/index.html accessed 9.2.2020

22. Mehra MR, Desai SS, Ruschitzka F, Patel AN. "Hydroxychloroquine or chloroquine with or without a macrolide for treatment of

COVID-19: a multinational registry analysis." *Lancet* May 22 2020 DOI:https://doi.org/10.1016/S0140-6736(20)31180-6 RETRACTED

23. Zachary Brennan. Fauci: Hydroxychloroquine not effective against coronavirus. *Politico* May 27 2020 https://www.politico.com/news/2020/05/27/fauci-hydroxychloroquine-not-effective-against-coronavirus-283980 accessed 9.2.2020

24. Large study finds drug touted by Trump is "not useful and may be harmful" for COVID-19 patients. *CBS News* May 22 2020 https://www.cbsnews.com/news/hydroxychloroquine-coronavirus-drug-study-not-helpful-harmful-heart-risks-trump/ accessed 9.2.2020

25. WHO halts hydroxychloroquine trial for coronavirus amid safety fears. *The Guardian* May 25 2020 https://www.theguardian.com/world/2020/may/25/who-world-health-organization-hydroxychloroquine-trial-trump-coronavirus-safety-fears accessed 9.2.2020

26. Melissa Davey. Questions raised over hydroxychloroquine study which caused WHO to halt trials for COVID-19. *The Guardian* May 27 2020 https://www.theguardian.com/science/2020/may/28/questions-raised-over-hydroxychloroquine-study-which-caused-who-to-halt-trials-for-covid-19 accessed 9.2.2020

27. IBID

28. Offord C. "Concerns Build Over Surgisphere's COVID-19 Dataset." *The Scientist* Jun 2 2020 https://www.the-scientist.com/news-opinion/concerns-build-about-surgisphere-corporations-dataset-67605 accessed 9.2.2020

29. IBID

30. Mehra MR, Desai SS, Kuy AR et al. "Cardiovascular disease, drug therapy, and mortality in Covid-19." *NEJM* May 1 2020 DOI: 10.1056/NEJMoa2007621 RETRACTED

31. Watson JA, Meral R, Price R, Simpson J on behalf of 174 signatories. An open letter to Mehra et al and The New England Journal of Medicine. https://zenodo.org/record/3873178#.XtmckdVKipr accessed 9.2.2020

32. James Watson on the behalf of 146 signatories. An open letter to Mehra et al and The Lancet. https://zenodo.orgJames /record/3864691#.XthfzTpKhPb

33. Retraction: "Hydroxychloroquine or chloroquine with or without a macrolide for treatment of COVID-19: a multinational registry analysis" https://www.thelancet.com/lancet/article/s0140673620313246

34. Retraction: Cardiovascular Disease, Drug Therapy, and Mortality in Covid-19. https://www.nejm.org/doi/full/10.1056/NEJMc2021225

35. Mulangu S, Dodd LE, Davey RT et al. "A Randomized, Controlled Trial of Ebola Virus Disease Therapeutics." *NEJM* 2019 Dec;381:2293-2303

36. Wang Y, Zhang D, Du G et al. "Remdesivir in adults with severe COVID-19: a randomised, double-blind, placebo-controlled, multicentre trial." *Lancet* 2020 May;395(10236):P1569-1578

37. Erika Edwards. Remdesivir shows promising results for coronavirus, Fauci says. *NBC News* April 29 2020 https://www.nbcnews.com/health/health-news/coronavirus-drug-remdesivir-shows-promise-large-trial-n1195171 accessed 9.2.2020

38. Adaptive COVID-19 Treatment Trial (ACTT). ClinicalTrials.gov https://archive.is/GRm9e accessed 9.2.2020

39. IBID

40. Appendix A, Table 2. COVID-19 Treatment Guidelines Panel Financial Disclosure for Companies Related to COVID-19 Treatment or Diagnostics. National Institutes of Health. https://www.covid19treatmentguidelines.nih.gov/panel-financial-disclosure/ accessed 9.2.2020

41. https://www.covid19treatmentguidelines.nih.gov/whats-new/ accessed 6.14.2020

42. Peter Mansell. Handle with care: changing endpoints in clinical trials. *PharmaTimes* April 19 2007 http://www.pharmatimes.com/news/handle_with_care_changing_endpoints_in_clinical_trials_989732#:~:text=If%20not%20appropriately%20evaluated%2C%20changes,lays%20it%20open%20to%20manipulation. Accessed 9.2.2020

43. Lev Facher. 6 takeaways from the Senate's surreal virtual hearing on the U.S. coronavirus response. *STAT* May 12 2020 https://www.statnews.com/2020/05/12/fauci-hearing-coronavirus-takeaways/ accessed 9.2.2020

CONSEQUENCES: THE CURE IS REALLY WORSE THAN THE DISEASE

ALMOST FROM the beginning of the COVID-19 response, many experts were warning that the cure was likely to be worse than the disease, particularly if it continued for an extended time. And it did. A couple of weeks to "flatten the curve" turned into a few more weeks and then months of restrictions with absolutely no end in sight. The emperors and empresses running the states and health officials even stopped putting forth a goal that if reached would put an end to their unconstitutional power grab.

There seemed to be no balancing of the potential harm from the virus versus the harm from policies inflicted on the public. Both governors and mayors continued to issue one crushing decree after another with complete disregard for the consequences of their decisions.

By the end of March 2020, people were stressed, with many begging for relief. By the end of the summer the carnage was almost unbearable. The economy had crashed, the unemployment rate was at an unprecedented high, homelessness was increasing, and the suicide and overdose rates had skyrocketed. The most vulnerable children had gone backwards academically, and nursing home patients were dying of neglect. Society as we knew it had evaporated and a growing number of people were realizing that the U.S. government, as well as the governments of other countries, had been overthrown by criminals. These criminals had declared themselves unaccountable to the public and demonstrated daily that they did not care about the impact of their decisions on their "subjects."

An honest evaluation of the consequences of the COVID-19 debacle must start with an examination of data showing which groups of people were at risk, and whether the measures taken were justified.

Impact Based on Age

The data are clear that the most vulnerable people during flu season every year are seriously ill or immunocompromised people of any age, and the elderly, and especially those who are sick or frail enough to be confined to nursing homes. This certainly turned out to be true for COVID-19.

On August 5, 2020, CDC Provisional COVID-19 Death Counts showed the following:

Total deaths from COVID-19 for all ages	142,164	
Total deaths age 85 years and older	45,845	32.2%
Total deaths age 75-84 years of age	37,495	26.4%
Total deaths age 65-74 years of age	29,870	21.0%
Total deaths age 55-64 years of age	17,583	12.4%[1]

These data clearly show that risk of death increases with age, and that most deaths (79%) were people age 65 and older. On the other hand, there were only 270 deaths in the U.S. from COVID-19 in people age 24 and younger.[2] The total deaths for people 54 and younger in the U.S. was 11,317.[3]

While every life lost is important, these numbers show that the heavy-handed and draconian measures implemented by government and health officials were egregiously off base and could not be justified.

Social Distancing

According to the Centers for Disease Control, social distancing, also referred to as "physical distancing," means staying at least 6 feet (about two arms' length) from anyone not living in your household. The CDC states that this must be done both inside and outdoors and must be practiced in combination with other strategies such as wearing masks, not touching one's face, and frequent handwashing.

The CDC's site makes several contradictory statements. First, it states that social distancing works only when combined with other strategies like wearing masks. If masks are effective, social distancing should not be necessary. Conversely, if social distancing is effective, then masks should not be necessary. The site also states that COVID-19 can live for hours or days on surfaces and that social distancing can limit opportunities to come in contact with contaminated surfaces. This makes no sense at all. A person can be 6 feet away from another person at the grocery store and still touch a theoretically "contaminated surface."

The CDC's site goes on to advise that people are best protected by not going anywhere at all. Shopping should be limited to only purchasing essential items and it's best to use delivery or curbside pickup. The message is clear. Stay as far away as possible from everyone except those you live with.

Social distancing has never been advised, let alone mandated, for the entire population for any period. Essentially what the government did was to declare by edict that it is conducting a clinical trial with hundreds of millions of subjects. Under normal circumstances, a clinical trial requires review before starting in order to limit potential damage from an experimental intervention. Like so many orders given by the government during this debacle, little consideration was given to the collateral damage likely to occur.

For many people, particularly those who live alone, instructions to practice social distancing resulted in isolation. There is considerable evidence of the negative effects of isolation. For example, a 2015 meta-analysis of 148 studies with 308,849 subjects showed that social isolation increased the risk of premature mortality by 29%. The researchers recommended that social relationship factors should be added to the list of other factors for early mortality such as smoking, diet, and exercise.[4]

Quarantining adults is associated with confusion, anger and post-traumatic stress disorder (PTSD). The impact is greater when increased duration, fear, and financial loss are added.[5]

Another analysis of 70 studies including over 3.4 million people showed that during seven years of follow-up, the likelihood of dying increased by 26% for those who reported feeling lonely, and 29% for those who were socially isolated. Surprisingly the risk of death was higher in those under age 65.[6]

While people of all ages are susceptible to the damaging effect of social isolation, the elderly likely suffer most of all. A report published in 2020 by the National Academies of Sciences, Engineering and Medicine concluded that social isolation was a risk factor for premature mortality in the elderly on par with high blood pressure, smoking or obesity. The report makes specific recommendations for identifying those at risk and the development of interventions to address isolation.[7]

For periods of time during the COVID-19 debacle, entire populations were essentially held under house arrest. Historically quarantine has only been reserved for the very sick, most likely because it is considered such an extreme measure and the risks are significant. A review of 24 studies that looked at the effect of quarantine during outbreaks of SARS, H1N1, Ebola and other infectious diseases since the early 2000s found that people who were quarantined experienced short-term and long-term mental health problems including stress, insomnia, emotional exhaustion, and an increased incidence of substance abuse.[8]

Studies have shown that the negative effects persist long after the quarantine is over. A follow-up study of 549 hospital workers in Beijing who were quarantined during the 2003 SARS outbreak showed that almost half reported alcohol abuse three years after the event.[9] Factors that increased the risk of psychological problems included quarantine lasting longer than 10 days (associated with increased risk of PTSD), and poor rationale for quarantine.

This effect is likely to impact children, too. A review of 80 studies showed that children and adolescents are more likely to experience high rates of depression and anxiety both during and after social distancing and isolation associated with school closings and home quarantine.[10]

According to psychiatrist Damir Huremovic, the health risks associated with social distancing are extremely concerning if the practice is extended beyond a few weeks. The collateral damage of this event includes economic recession, unemployment and overall uncertainty, and combined with social distancing these can trigger "unpredictable and widespread health challenges." He went on to say, "I sincerely hope we do not get to this stage."[11]

During the debacle, people were encouraged to stay in touch by phone, Zoom, Skype and the use of other technology. But this is not a substitute for in-person contact, holding hands, or hugs. It is normal for humans to want contact with other humans. We are, by nature, social creatures.

It is fair to ask whether social distancing is effective given the known risks associated with this practice. As mentioned earlier, there has been no research on the use of this practice in entire populations. However, there are two relevant and very public reports that merit coverage.

In May, New York Governor Andrew Cuomo reported that most hospitalizations in his state were people sheltering at home and who had almost no contact with the outside world. The data was drawn from 113 hospitals and showed that 66% of patients had self-quarantined, as compared to 18% who were nursing home patients. Claiming shock about the data, he said, "This is a surprise: Overwhelmingly, the people were at home," he added. "We thought maybe they were taking public transportation, and we've taken special precautions on public transportation, but actually no, because these people were literally at home."[12] These findings did not cause him to change his policies, however.

A 60-page analysis conducted after the massive protests following the death of George Floyd showed that there was no spike in cases or deaths due to the protests themselves. The authors noted that there may be several mitigating factors, including the age of the protesters, and business closures in the general vicinity of the activity.[13]

The Bottom Line

It appears that like almost everything else associated with COVID-19, efforts to "protect" people were ill-conceived and not based on science. In addition to the research presented herein, the fact that lockdowns, restrictions, and school closures continued after months of "following the science" from the "experts" provides evidence that wearing masks, social distancing and quarantine were failures.

"Protecting" the Vulnerable

Even the lockdowns of nursing homes seem outrageous when looking back at the impact of residents. Almost everywhere, visitors, including family members, were prohibited at nursing homes and extended care facilities. Employees were tested every day to reduce the risk of infection. But the strategy was miserably ineffective. On August 5, 2020, the Ohio Department of Health reported that 2060 patients had died from COVID-19 in long-term care facilities.[14] Total deaths from COVID-19 on that date were 3668.[15] In other words, Ohio nursing home patients were prohibited from having interaction with anyone other than staff and still 56% of all deaths from COVID-19 took place in these facilities. Even now, long-term care residents can only see visitors outdoors, residents and visitors must wear masks, and social distancing is required. Visitations must be scheduled in advance and if it rains, most visits are postponed.

Summertime in Ohio is hot, which makes outdoor visits challenging for some older people. The mask requirement can make these visits almost unbearable. Stroke victims have difficulty communicating under the best of circumstances and masks make it impossible. Those who do not hear well don't get much out of visits which require 6 feet between residents and their guests. Many frustrated people have stated that the Emperor DeWine thinks it is ok for long-term care residents to get COVID-19 from strangers but not from family members.

The consequences of the isolation and restrictions? Unbearable. Most of the staff in long-term care facilities are caring people who want to offer the best care to their patients. Still anyone who has had a family member in one of these facilities reports that the constant presence of family members and friends helps to keep the staff accountable. Both family and staff agree that regular visits are extremely important for the mental, cognitive, and physical health of patients. These facilities were not designed with the intention of staff providing companionship, interaction, or extras like baked goods, flowers, books, and DVDs.

When visits were cut off, it did not take long for care to degenerate, and for patients to start suffering from loneliness, boredom, and mental and physical decline.

Here are just a few heart-breaking stories.

I'm in Florida and the rest of the family lives in Quebec, Canada. Last Monday my sister called me crying and very distraught. She'd been called by the retirement home where my father lived. They told her they had found my father non-responsive and rushed him to the hospital. She wasn't allowed to go to the hospital to be with him yet but they asked her if she could come clean his room while he wasn't there. We pay extra for that cleaning service, mind you.

When she arrived, she found food and beverages all over his room, much of it with mold on it, stinking and rotting. She found his dirty Depends diapers everywhere. Under the bed, the couch and bathroom. She spent half the day cleaning this mess while balling her eyes out.

The following day, the doctor called her to the say there was nothing they could do and since he was terminal, now she could come visit. She was never able to have a coherent conversation with him the whole week and he passed a week later.

Since this whole nonsense started, all she could do was stand in the parking lot while he was peaking his head out his 3rd story window. She'd ask him "how's everything?" And he'd respond "Fine."

She couldn't see the mess in his room. He wasn't eating and was withering away. Something she would have noticed if able to visit him as usual.

The residents in this home went from a routine of going to the cafeteria 3x day, eating their favorite food. My father had no teeth so it was usually soft eggs and a juice, soup and mashed potatoes etc. He socialized with friends, spent time on the patio

And then one day, he was imprisoned, confined to a small room for months.

My father got too lonely to cope with this and just gave up and stopped eating. AND they were bringing him coffee that he doesn't drink in the first place and fruits he couldn't chew.

*   *   *

I wanted to let you know that I am sick with worry than my 91-year-old mom, who is in an assisted living facility in Oakwood, Ohio, might succumb to the quarantine rather than the virus. On December 15, 2019, we lost my father. My parents were married 66 years. We moved my mom to a pricier facility with the idea of having more activities and chances to interact with others

Then the quarantine. None of her four kids can visit her. She has been hysterical, and the facility does not like it. Consequently they load her up on drugs. I just had a Zoom conference with her and she couldn't communicate. She arrived at this new facility with slight dementia. The loneliness and isolation are killing her. I am writing to you because I want her voice heard.

*   *   *

My husband's Aunt and Uncle, who are in their late 80's, reached a point where they were no longer able to care for themselves and so they checked themselves into an assisted living center about a year ago in The Dalles, Oregon, where they were born and raised. He had been a career minister and, besides being a Sunday school teacher, his wife volunteered selflessly in every way possible helping out the church and community. Both were very sociable, beautiful people with countless friends who loved them dearly.

When the lockdowns ensued in March, their health started going downhill rapidly. Of course no one was allowed to visit them, but when his Aunt's health failed further, they took her away to a separate intensive care ward in the same building, and not even her husband of 65 years was allowed to visit her. There has not been a single case reported in the facility. Of course they would argue that's because they did such a good job locking everyone up :(. To date, there have only been 3 deaths in a county of 30,000, and who knows if those 3 were even legit.

Two months later, his Aunt died (NOT from covid!), and the family wanted to have some sort of funeral for her, and involve his Uncle in some way. The only solution allowed was to hold a small service of immediate family only, in the parking lot of the care center where the Uncle, who is extremely hard of hearing, could watch through a window if we all wore masks. So the day came, and as luck would have it, the weather did not cooperate. We all gathered outside of the window, in a torrential downpour, getting absolutely drenched (including my husband's 90 year old mother) trying to hold a service, while the Uncle sat in his wheelchair behind the window where he could watch. There he sat, looking out at a bunch of masked faces, unable to hear, unable to be comforted, and we could all see him crying like a little baby and were helpless to console him in any way. He was not allowed to go with us to the burial site, so we tried saying our good bye's by waving at him through the window and I expect that will be the last we ever see of this kind and gentle Godly man. I'm just relieved that my 98-year-old mother died two years ago with all 8 of her kids at her bedside, as it should be. What kind of cold, heart-less, cowardly, unconscionable people do this? Great moments in public health, indeed.

Perhaps saddest of all was a news story featuring 52 nursing home patients in Gatesville Texas who were photographed hold-ing signs asking for people to become pen pals.[16]

The criminals and despots in charge of our lives seem un-concerned by the consequences of their decisions, which clearly did not prevent death from COVID-19 in long-term care facilities. Data from most states and even most other countries was similar to Ohio. Most deaths were in elderly people, and a significant per-centage of those who died were locked down in nursing homes.

According to psychiatrist Dr. Peter Breggin, the isolation of seniors "...was contrary to every principle of caring for the elder-ly. There is no controversy about the best way to help the elderly with their overall health, cognitive and emotional problems, or dementia. Keeping them in close touch with the people who love them while providing maximum autonomy and opportunity for a degree of normal functioning is critical to maintaining the mental and physical function of these fragile humans. The restrictions imposed by the lockdown on nursing homes was devastating to the morale and the health of the patients, destroying both quality of life and life itself."[17]

Impact on Children and Adolescents

Lockdowns and school closures were particularly difficult for children. According to the American Institute for Economic

Research (AIER), closing schools in March was, essentially, a grand, unethical social experiment designed by supposed infectious disease experts. The AIER says, "We consider the experiment to be unethical because there has been no informed consent, either from parents, children, or even our legislative representatives. The bottom line is that our children's future and the quality of their lives have been sacrificed to conduct this experiment."[18]

Schools had little time to convert to "schooling at home," and the challenges in making this conversion were almost unsurmountable. According to an April 7, 2020 article in the *New York Times*, in rural communities many children did not have internet access. In these areas, educators reported that students and parents just dropped out of touch and were not available by phone or email. Absences were very common in low-income school districts. According to Michael Cassidy, Executive director of the Council of the Great City Schools "The dramatic split promises to further deepen the typical academic achievement gaps between poor, middle-class and wealthy students and unfinished learning will be a serious issue that could have implications for years."

Eric Gordon, CEO of the Cleveland Metropolitan School District reported that 30-40% of students did not have access to internet. A teacher in that district reported that most of her students' parents do not speak English.

A school district in Minford Ohio distributed laptops and work packets to students. But Mari Applegate, school psychologist, reported that regardless of whether or not the students can log in or turn in assignments they will be passed on to the next grade since it is not their fault and "cannot be held accountable."[19]

An analysis of 800,000 students conducted by researchers at Brown and Harvard Universities determined that student progress in math decreased by about half in poor zip codes, and one third in middle income zip codes. Kids in high income zip codes were not affected. They estimate that the average student will fall behind by 7 months, 9 months for Latinos and 10 months for black children. The hardest hit may be rural areas, since only 27% of schools in these areas required any instruction at all while the schools were closed.[20]

In addition to falling behind academically, keeping kids out of school resulted in lack of social and emotional development because they were deprived of play, sports and other activities. Autistic and special needs children were hurt most, since their routines were interrupted, and they had little to no access to the specialized help that they required.

According to Shelley Allwang, program manager at the National Center for Missing and Exploited Children (NCMEC) the COVID-19 response resulted in a significant increase in reports of child abuse. In April 2019, NCMEC received about one million reports. During the month of April 2020, 4.1 million reports of child abuse and exploitation were reported.

The staff attributed the increased reports to bad actors who were taking advantage of children who were out of school and at home, along with parents who were overwhelmed with schooling at home while trying to work. Kids suddenly stopped spending time outside and playing with their friends, and spending more time online than ever before, which created more opportunities for exploitation. Another issue was assignments from teachers that involved internet searches. The searches were broad and had the potential to result in visiting inappropriate sites, or alerting bad actors patrolling online that the child was online and available.[21]

Adolescents missed important rites of passage, like the school prom and graduation. Their college educations were interrupted, and they were sent home with nothing to do. For athletes, the consequences might be particularly punishing since college sports are often the ticket to lucrative professional careers.

Here are a few stories demonstrating these outcomes:

I am so angry. My son was enjoying his junior year of college abroad in London. First his friends at the university were ordered back one by one to their home states, some states which had higher rates of the virus than in London. I kept telling him to stay because there was more risk (although still extremely low) of flying back then just staying put. Eventually within the span of a couple weeks, he was practically the only one left. His school was open so I still told him to stay. Then the school eventually went online, but he was still allowed to stay. Then the school closed and he was basically kicked out. Now he's home and sits in his room all day except for taking a run or the dog for a walk (which I'm worried will eventually be banned also.)

I have another son in college forced to come home and he sits in his room all day also. I'm more worried about their mental health at this point. They are both well adjusted, happy, social kids, so I hope this doesn't affect them too negatively. I seriously worry this will trigger some sort of mental instability. These young people should be out living their lives!

*   *   *

This mishandled situation has significantly affected our family. We have 2 teenage sons, one is a sophomore and one is a senior. At the beginning of March our senior had a part time job and was successfully completing his senior year in high school and eagerly looking forward to the end of year senior activities. Now he and the other classes of 2020 are missing out on those milestone activities and graduation. Now he isn't working and is home doing distance learning. He has been working on his Boy Scout Eagle Project process for the last year and a half. Thankfully his Eagle Scout service project at a county park, which had been scheduled months in advance, was a week and a half before the shelter in place orders were issued here in the middle of March.

Our younger son, a sophomore, was in the process of starting his Eagle Scout project but his project has been postponed indefinitely due to the shelter in place order and cancellation of non-essential activities and gatherings here. Future schooling (college and high school) is completely up in the air for our sons, based on whether or not schools reopen or continue with online and distance learning.

*   *   *

My niece is traumatized by missing all her high school senior year activities she looked forward to for four years, not to mention she still had to meet the deadline to choose a college while not being allowed to visit any of the campuses.

The same niece's teenaged boyfriend who suffers from depression had suddenly gotten so bad the past month in isolation that he said he didn't think he would make it another month and would kill himself. She grew so concerned she contacted his mother, and he cannot get an appointment with a therapist to get medication for a few more weeks.

Suicides and Overdose Deaths

The increase in drug overdoses and suicides started shortly after the lockdowns began. This is not difficult to understand since the effect of unemployment, business failure, isolation, financial insecurity and other consequences of the lockdown are well-known.

Doctors at John Muir Medical Center in Walnut Creek reported on May 21, 2020 that there were more deaths by suicide during the quarantine than deaths from COVID-19. The head of trauma, Dr. Mike deBoisblanc stated it was time to end the shelter-in-place order because it was clear that the hospitals were

not overwhelmed, there were adequate resources to take care of
COVID patients, and the rest of the community was suffering.
"We've never seen numbers like this, in such a short period of
time," he said. "I mean we've seen a year's worth of suicide at-
tempts in the last four weeks."

Kacey Hansen has worked as a trauma nurse at John Muir
Medical Center for almost 33 years and expressed concern because not only was the facility dealing with more suicide at-
tempts, they were not able to save as many patients as usual.[22]

Robert London MD, a psychiatrist, noted that the country
was experiencing "a national epidemic of trauma," which he
described as "a clinical picture of PTSD." He stated that isolation…is both painful and stressful," and cautioned that worry
about many things, including family, finances and work is overwhelming for millions of people. In addition to social distancing, for many people there is no work, people cannot spend time
with people they care about, there is no recreation, no shopping,
no normalcy.

London expressed concern about people experiencing
nightmares, anxiety, and insomnia, and reports that people were
stocking up on guns and ammunition.[23]

Some have spoken out against the tragedies inflicted on our
youth. Here is one example:

I'm exhausted.

I'm tired of the garbage on FB and mask shaming and
self-righteous sanctimony spewed by every noodle out

there who thinks they are right and if you disagree, you are not only wrong but *dangerous*.

The 15-year-old son I've never met of people we knew during our Seminary years killed. him. self. the other evening. At home. Feet away from his pastor father and mother. No warning. Their only child. He couldn't process the "new normal" and the constant drumbeat of negativity. The loss of everything a normal child needs in his life.

Shame. On. Us.

That we have allowed this to happen. That we have sat by complicit in the erosion of liberties and the decimation of our economy. That we have excused the behaviors that have left adults and CHILDREN so bereft that suicide can happen right in their home. All because of a bad flu. There. I said it. THE FLU.

Hate me? Think I'm wrong? Crazy conspiracy theorist? Don't bother commenting. Two can play at this. You cannot justify a total lack of compassion and empathy at this point while demanding compliance. Door's over there, see yourself out. I'm done defending myself. I'm done being castigated for daring to think differently. There is no tolerance. No discussion. No agree to disagree. In a world where the EXPERTS can't agree, and have changed their stance a hundred times, *I'M WRONG* when I don't immediately buy the science du jour. And who pointed it out? My 15yo son. Who spends 8 hours a day in his room, on the internet, "doing school". Not with friends. Not at the rink. Not at school. He said "When did it become only one thought process was acceptable?" NEVER, Lex. It should NEVER be that way.

I thought my charge was to protect my kids from a virus. The virus doesn't scare me anywhere near as much as this insidious darkness falling over society. Where neighbors are tattling on each other. Actually MAD that other people are not living in fear. "Defying orders". Embracing masks which make everyone look angry and hide expression, furthering the feeling that everyone is suspicious and dangerous. I can't wear a mask, and the amazing thing I've found is people are GLAD TO SEE MY FACE. The like to see my smile! They miss human contact and they know they can converse with a mask-less person because we aren't going to freak out. I haven't encountered one single cashier who wasn't delightful and pleasant (except Menards, and they need Jesus over there, for sure).

No, my charge is to protect them from the Biblical levels of darkness falling. Their entire world has been damaged, and to believe that the love and protection of their parent is enough to save them is naive. Their world is far bigger than me, and it should be. But it has been taken from them and there is no hope being offered by ANYONE in authority that they will get it back. Nope, instead there is the constant drumbeat of "new normal". That discourages *ME* and I've got the maturity and life experience to process that better than our children.

So...I'm not participating. I'm not excusing any of it. I'm not playing the middle. I'm not trying to be "balanced". I'm done. I refuse to sit by and just mourn, from a distance, no less, for a CHILD who despite being raised in the church, by devoted and loving parents, and showered with the love of Christ, still was so hopeless he was compelled to take his life at 15.

Shove your masks. Screw your fears. I'm done participating. I'll be in church, same row as ever, no masks, hugging all takers. I will not contribute to the darkness.

Shame. On. Us

The Inhumanity of it All

During a five-month period, we received thousands of horrible stories about horrible and inhumane treatment and which reflected a general disregard for human life. Many of these people reported that they had contacted, called and written letters to government officials and were either ignored or treated badly. In other words, reports of harm and pleas for mercy fell on deaf ears. Some people who answered the phone casually mentioned that "a lot of people were complaining" but appeared to have become immune to hearing about suffering people and seemed not to care.

Here are a couple examples of these stories:

On Saturday a friend and client passed away. I'm so sad and angry. Not that it's ever a good time to die or go through the many difficulties of cancer, but doing it during this COVID19 "pandemic" is horrific. She turned 59 on April 8th. Her mom and sister had booked plane tickets at the beginning of March to visit her for her birthday. Due to COVID, they were not able to visit her before she passed away. On April 13th the pain she was experiencing from ascites was so bad her daughter took her to the ER. She stayed in the hospital ALONE for 4 days and was then released to hospice care. At the end of March, when she really started going downhill, I asked her if there was something she really wanted to do. She wanted to go to the beach. I was unable to grant her wish because the beaches were closed due to COVID. There are so many more atrocities I witnessed her go through during her cancer treatments, but I can't continue typing right now. I'm just so furious with the whole medical establishment and overcome with sadness.

*   *   *

My husband and I have 3 kids who still currently live at home with us. The oldest is my son from a previous marriage, he is 22 and is intellectually disabled. He has suffered the most from all of this tyrannical behavior, so my letter to you is mainly about him. He is considered high functioning autistic. He is able to do things for himself and be fairly independent, however he will never live on his own independently.

He will never drive, so therefore he takes the county transit in order to go to and from work. He has worked at his job for over 3 years, he loves it there and his fellow employees are very kind to him. Customers love him and some come in just to see him and his big smile.

He considers himself famous because he is very involved in Special Olympics and went to the 2018 USA Games and last year we went to Abu Dhabi for the World Games to watch him compete and bring home 4 gold medals in Powerlifting. He was on cloud nine and living up his best life.

But lately, that smile has disappeared, it has been replaced with anger and frustration, uncertainty. He is not the same person he was at the beginning of this year and I completely blame it on our state governor, whom I don't even like saying his name anymore at this point, mr dictator dewine. Along with his former sidekick, Acton, they have caused my son severe mental anguish that I fear will affect him long term. We have come so far with my son, and now all his successes are failing him.

My son went to his local gym 3 days every week so when the dictator of Ohio closed gyms, this was traumatizing to say the least. Fortunately we have gym equipment here at home, so that became his only outlet. It still wasn't the same though overall. He still has not returned to the gym due to the strict guidelines that have been ordered in order for gyms to even be open. He wouldn't have the time to follow everything and still get his workout in before work. He is also afraid he will not follow as well as is being demanded and therefore be kicked out of the gym. Along with the fact that they make them sign in, which I am not willing for him to do because of the whole contact tracing mumbo jumbo. He ended up almost, call it what you will, having a mental breakdown over all this.

He was off work for about a month, with no income because his place of employment never shut down or closed. He just couldn't even function at work so there was no point in him going there. He couldn't go to acupuncture because they were closed down, which is vital to him and a vital balance for his mental health, so that on top of everything else, he was like a ticking time bomb. To see your grown 22 year old son breaking down, sobbing...it's not something I even knew how to handle and I know him best.

Once things started opening back up again and it seemed more relaxed for him and he could go back to acupuncture, he then returned to work and I was seeing his smile again. He was still frustrated at times because of the mandates the health departments were now enforcing and would come home and need to tell me everything they were making him do. We would talk it over and then he was fine.

Then the mask thing started intensifying. He was seeing people on social media shaming others for not masking up and implying those that don't do it, do not care for others. I told him to quit looking at it, those people are not doctors and they don't know everything. He said but I do care about people, I just can't handle having that on my face. He is a kind soul who would do anything to help others, truly unique. He is a good boy :) He kind of has a tick type thing where he tends to rub his fists on his face, not all the time, just some. He is careful and keeps his hands washed and tries to control touching his face. Honestly, I believe our whole family already had this virus back in February, but that is not important to any of our rulers.

Back to the masks...He obviously is exempt but that did not keep people from making comments to him. We had a plan and it was just that he was to respond by saying it was none of their business and smile, walk away. Thankfully, this did not happen much. Now that licking county was moved to red level and masks were mandated last week because we had like 14 positive cases added to our faulty total of like 400 cases out of nearly 180,000 people, I was more concerned for my son and what might be said to him. People can be cruel and this whole masking has gone way out of control. No one seems to care that there are actually people who cannot tolerate or physically can't wear one. Our entire family are not mask wearers and we are doing all we can to avoid that entirely.

My son was encountered by his transit driver just yesterday, telling him if he didn't wear a mask that he would be kicked off and no longer permitted to ride. Not a good thing to tell my son at all, he was furious and called me at work. He was to the point of almost crying. I called the county transit service and got it straightened out and noted on his file that he is indeed exempt. The kicker was at first, they were going to require a doctor note to excuse him from wearing one. I was like what? That was not mentioned in the health departments standards at all. They hem hawed around and I told them he is exempt from wearing a mask along with exempt from having to get a doctors' note because he is on a waiver through the County DD and they have transportation provided by county transit as part of his ISP plan and the DD pays for it as part of his plan. That settled it and I was glad to have it resolved. I was just somewhat still taken back by the fact of the doctor note request, where did this come from?? People will be required to now prove why they can't mask up??

Sadly, my son thinks our world is ending now and he is back to his state of mental breakdown mode once more. So to all this, thank you so much dictator dewine for ruining my happy, smiling, somewhat carefree son who had everything going for him - you basically took it all away in a matter of several months. He was set to compete in more competitions this year and obviously participate in the Summer Special Olympic games, which was canceled due to all of this mess. The last time my son competed was at the Arnold Sports Festival, which thankfully that happened just in time before the tyrants took total control of that situation. My son misses so much of what has made him who he is, that he has actually lost himself in the midst of everything. He is even having trouble identifying with himself and can't think straight, he is confused easier. He hasn't seen his special olympic team mates since, I honestly can't remember when.

In closing, now with the episode that happened yesterday with transit and other things that have been mentioned at his work and his mental state, he is going to have to take more time off work. My son is at the breaking point and what dewine has done and is doing..I hold him ultimately responsible for. He has made my son with his pandering rules and regulations into someone who is not happy and can't handle the day to day. Yesterday was the first day since my son first got a job, in like 3 plus years, he has begged me to not have to go to work. This guy has loved his job, a totally dedicated employee that any employer would be happy to have. He is a hard worker and now his face is long and tired, he is mentally exhausted. I have it in my mind to speak to an attorney friend, because my son has lost income over this and mental well being. I have never seen someone have a mental breakdown and I don't care to either, but it has come to that point. Everyday life is actually harming my son!!! Who would have ever thought that.

I just wanted to share my story with you as far as the mental toll this is taking on the intellectually disabled and how basically criminal this all is. Our state government is not taking any of this into consideration at all, its harmful and why so many suicides have occurred. DeWine would rather have someone get on and talk about things to do for your mental health, he forgets about the ones that do not understand what that even means. I hope and pray for better days ahead, but the path this is going is dim and I don't see it getting any brighter any time soon...at least not until the election perhaps ;)

The Economy in Freefall

In February 2020, the unemployment rate in the United States was 3.5%. In April of 2020, due to the government response and shutdown, 20.5 million jobs were lost.[24] The numbers continued to rise throughout May, and as of May 8, the unemployment rate had risen to 14.7%, according to the Bureau of Labor Statistics.[25] The United States had a thriving economy, a thriving stock market, and companies were reporting record earnings. Almost overnight, millions of people saw their businesses evaporate, their life savings depleted, and a significant percentage were thrust into poverty.

Government officials felt compelled to do something, and tens of millions of now financially insecure people were demanding assistance. The government's response? Congress passed a $2.4 trillion dollar stimulus bill designed to help Americans to survive.

This stimulus bill pushed The United States' national debt to well above $25 trillion, with China continuing to own a large percentage of it. This move, which essentially meant the government was printing money it did not have, will most certainly result in a devaluation of U.S. currency, and will ultimately cause inflation to rise.[26] In fact, many economists, including Martin Hutchinson, a well-known author and market analyst in the industry, believes that by early 2022, this will lead to double digit inflation rates.[27]

Qualifications for the check were based on most recent tax returns (either 2018 or 2019) and any person whose adjusted gross income did not exceed $99,000 received a check.[28] This included people who were still employed, many of whom were happy to receive a check from the government, but did not need it. On the contrary, it is doubtful that the assistance was meaningful to most households that did need the money. Other than feeling better for a short time, how helpful is a $1200 check to the average person who has lost some or all of his household income? Most likely not much.

Most middle-income Americans tend to live on what they make. A family making $50,000 per year tends to live in a residence and have other expenses that are affordable within that range. This is also true for individuals and families with a $100,000 income or more, and these individuals and families were excluded from the stimulus check. For example, a restaurant owner with an Adjusted Gross Income of $100,000 in 2018 or 2019 was not eligible to receive a check, even though his income currently may have been reduced by 50% or even dropped to zero due to lockdown restrictions.

Another $3 trillion dollar bill introduced by Congress on May 12, 2020 was no better.

Included in the "stimulus" was bailout money for the already failing and mismanaged postal service, and "election assistance," which included dropping the requirement for identification to vote.[29] This, of course, will have the effect of making it easy for illegal immigrants to vote. It offered little in the way of relief for the tens of millions of people who had lost their jobs or whose businesses remained closed. It had not passed as of mid-August.

Art Laffer, a former economic advisor to the Reagan administration who was opposed to this approach to resolving the economic crisis the government created said, "Whenever people make decisions, when they are either panicked or drunk, the consequences are rarely attractive and that especially goes for politicians, so I thought Trump's proposal of a tax cut on payrolls was great, but that's about all I would suggest him doing."[30]

As of the time that this book was being finished, there were no meaningful proposals from Congress, or anyone else for that matter, that had the potential for addressing the coming explosion of joblessness, financial devastation, homelessness, food insecurity and other related consequences of the draconian measures the governors of various states had caused. And the long-term consequences for both the economy and individual businesses will most likely continue to get worse.

Gross Domestic Product (GDP) is often the primary metric used to determine the state of the U.S. economy. A recession is typically defined as two consecutive quarters of negative GDP. Given the unemployment rate and business shutdowns, many economists have made dire predictions for the U.S. economy. Goldman Sachs predicts an annualized decrease in GDP for 2020 of 34%. Deutsche Bank predicts 33%. JPMorgan's prediction is grimmer at 40%.[31]

Some economists are optimistic that the economy will bounce back toward the end of the year but given the fact that it took several years to build the economy from the last disaster in 2008, this is not likely. Many states implemented a phase-in plan for businesses to reopen, but many remain closed, or are forced to operate at limited capacity, like restaurants; or with extreme restrictions like gyms. Even for those businesses that fully open, sales are down since customers have been frightened by the ginned-up hysteria over "cases" and are afraid to shop or remain unemployed and thus have little money to spend.

A discussion of the post-lockdown economy would not be complete without returning to China and the CCP. We may never know if the release of the virus was deliberate or accidental, but the CCP was certainly not going to let an opportunity go to waste. The CCP withheld information, in collusion with the corrupt World Health Organization (remember that Tedros re-appointed President Xi's wife, Peng Liyuan to another 2-year term as "goodwill ambassador to the WHO) which allowed the virus to spread while most in the world were unaware of its existence.[32] While the data make it clear that COVID-19 never qualified for pandemic status, the WHO declared it as such, which is what triggered the closing down of the entire global economy, including that of the U.S.

Knowing full well that troubling economic times have historically created difficulties for an incumbent President to get re-elected, the CCP had a powerful incentive to play a role in these events. Joseph Bosco, former China Country Director for the Department of Defense stated, "Xi might well have asked his colleagues: Who will rid me of this troublesome president?" Suddenly, thanks to the export of China's virus, Trump's reelection prospects seem considerably less favorable than they did just a few months ago. A return to a more accommodating U.S. China policy with a new president seems disturbingly more likely.[33]

In fact, right on cue some politicians in Washington began calling for the removal of the tariffs, and if they are lifted, the CCP will have free range to saturate the U.S. market with its products again. Michael Wessel, a member of the U.S.-China Economic and Security Review Commission stated "As our steel and other manufacturers all suffer, when the bottom hits, China is poised to come back in, China is now looking at ways of taking advantage of everyone else's suffering."[34]

The Chinese haven't exactly been quiet about their delight about recent events. Han Jian, of the Chinese Academy of Sciences and director of the Ministry of Civil Affairs for the China Industrial Economics Association, and who received his doctorate at our own Johns Hopkins University, said on March 4, 2020: "It is possible to turn the crisis into an opportunity — to increase the trust and the dependence of all countries around the world of 'Made in China'."[35]

The Impact on Businesses

It is almost impossible to describe the full extent of the devastation and destruction the lockdowns had and will continue to have on businesses. People literally watched their life's work destroyed within a few short weeks to months. Both frustration and anger are palpable in the following accounts.

I am mad as hell.

The intent of this statement is to share with you and all others involved, that the Regulatory Taking without compensation of my businesses in the emergency period, has been misused and abused by Governor Jay Inslee of Washington State. The original intent of the 2 week Stay at Home order was to slow the spread of the Covid-19 virus from China. Judging by the infection numbers, death rates of actual Covid patients and the counties affected, Jay Inslee has intentionally killed my livelihood in Chelan County and I demand it stops now!

We are in the 11th week of a 2-week shut down. I am mad as hell that I have had to take 2 loans against my business that I have painstakingly built 7-days a week for the last 24 years. I am mad as hell that at 50 years old, the term on the SBA loan puts the payoff on my 80th birthday. These new loans may not even save my businesses, because the re-opening of our County has been batted around as a toy by our Governor and King, Inslee. This is not a game, this is life and death for us and I am mad as hell.

When asked to be a good citizen and "flatten the curve" and then I hear "we are all in this together" my blood boils because I know its political bull-excrement at this point and I am mad as hell. We are not in this together, if we were, there would be a remedy post haste. I have laid off all 63 employees that have families, mortgages, rents, diapers, food and the list goes on. The extended actions of Jay Inslee have put their lives in danger. How dare you Jay?

His orders have crushed my $3 million a year business with a payroll of more than $1.4 million because we are deemed "non-essential". By scaring our communities to the point of endless isolation, the future of my business is also in jeopardy.

Since we were unable to open this past weekend for Memorial Day and begin to dig out of this mess, it was a slap in the face of every soldier that laid down his life for the freedoms we celebrate. If I lose my life's work to Jay Inslee's ignorance and arrogance, by no damaging act of my own, there will be restitution due.

*   *   *

Marie Ann Longlade School of Dance has been my mother's business since 1965. Yes, 55 years in the business of sharing the love of dance. A female entrepreneur that rode out the recessions and made sure, no matter what, that "The show must go on"

For the first time, in all those years, the catch phrase of "The show MUST go on." FAILED. The studio has been hosting remote Zoom classes, since the end of March break. Parents paid a $100 deposit for their child's recital costume for each class the child was registered for. Some costumes were made already in preparation of the International Dance Educator's of America and Canadian Dance Teacher's Association Spring. These competitions were obviously cancelled.

Her 55th Spring Melody Recital with a BIG surprise celebration including of decades of past students returning to the stage in honour of this tenacious woman WAS CANCELLED.

*   *   *

As a result of the FEAR that Trudeau has instilled, and the crushing cancellations throughout 2020.... my mother, age 77 with COPD, and my step-dad, age 66, with Parkinson's have decided to give up one of their vehicles. They made this choice in light of the NEW NORM. They are convinced my mom will need to restrict her lifestyle forever more.

Not only has this destroyed her business, but it has crushed her drive to live. My step-dad has been observed crawling in his hands and knees to get across a room... with tools in tow as he is constantly, building repairing, and creating things. Parkinson's and COPD never stopped them. But COVID has held them hostage in their home since Feb 27th.

My mom had pneumonia in December. They found nodules on her CT that were highly suspicious of Cancer. They repeated the CT IN Jan. Still there. They took biopsies, but the results were negative. The Specialist told us she was certain it was a false negative and reordered a biopsy. She endured 12 hours in hospital in Feb due to a collapsed lung. The results were still inconclusive.

Then all hospitals shut down except for severe COVID patients. Sooooooo, nothing has been done since. My mom thinks she has cancer and is waiting to die, paralyzed to live life as she once had.

Do you think this NEW NORM of isolation for the rest of their lives is healthy? I am so sad that their fear has totally stripped them of living a full life. They are being robbed of life!

*   *   *

Shelli and I did not post this to offend anyone, from our perspective this was very good news for our employees and our business. We have worked tirelessly for the past 20 years building first our ice sculpting business and then the restaurant. Then in a very short amount of time all this work has collapsed around us without our doing. It is very hard to see all that we have worked for, the very future of our children and ourselves disappear.

We grieve for those of you who have lost loved ones thru this illness and we continue to follow the guidelines with the exception of wearing a mask.

We do greatly care about our patrons and our staff. We believe everyone has a choice. We offer both indoor outdoor dining and our staff each has a mask in his/her pocket and will wear it at your table if you ask.

Our staff tried to wear them but at the rate they rush around and the summer heat they were having difficulties breathing and felt faint.

At this point though for us it is very difficult to see this as not being politically motivated. We have faced 4 months of either forced closure or lowered abilities to have our business open while so many others were not financially burdened. We are facing unrealistic minimum wage and unemployment issues that threaten the very fabric of our business and therefore our lives. With all the government has done to the hospitality industry we have not been able to catch up and make ends meet. We are not being taken care of by the government. Shelli and I did not collect unemployment. Nor have we received all the promised grants for the business. We are 500,000 in debt for this business and prior to the forced shut down we were a financially sound business. We had to make a choice to open without restrictions so we could try to rescue our livelihood and keep the 35 individuals that work for us employed as well as ourselves.

This is in no way a political stance we are just fighting for our livelihood
What made business owners even more furious was the backlash from government officials and citizens who had been brainwashed by the media. These people accused anyone who

was concerned about their livelihood as being a heartless capitalist who cared only about profits. Business owners who were justifiably panicking about their economic futures had to endure accusations from self-righteous and indignant individuals that they were selfish and had no regard for human health.

In fact, the selfish ones were those who seemed to lack any understanding about how the world works. The economy is what sustains human life. For many people, their business is their life's work. Business owners, along with their employees, often are driven by purpose – to better the lives of one another, their families, and their customers. All businesses are essential – to the people who own them and who are employed by them. Businesses allow people to take care of and feed their families and to contribute to their communities. Communities rely on businesses for their tax base, funding of arts organizations, museums, and culture. To suggest that businesses are not important is preposterous. And further, to think that commerce could all be indefinitely suspended and then reconstituted at some later date is insane.

Worldwide Impact

According to David Bealsey, head of the World Food Programme, the world is now at risk of families "of biblical proportions" going hungry as a result of actions taken in response to COVID-19. He noted that many parts of East Africa and South Asia already had severe food shortages caused by numerous factors such as drought and insect infestation. He referred to the developing famine as the worst humanitarian catastrophe since WWII.

According to Beasley, the crisis most affected millions of people who were "already hanging by a thread," and is particularly devastating for those "...who can only eat if they earn a wage."[36]

According to a report by Oxfam, the COVID-19 response is likely to throw another half a billion people into poverty. This would be the first time since 1990 that poverty increased and could be severe enough to put some countries back to where they were three decades ago.[37]

Worldwide, the COVID-19 debacle will likely result in an additional 1.4 million deaths from tuberculosis, 500,000 additional deaths from HIV, and 385,000 additional deaths from malaria. The reason is that delivery of services to sick people have been disrupted, diagnostic testing has been delayed, travel to receive treatment is difficult and often impossible, and access to medications is limited.[38]

Indeed, the cure was far worse than the disease. And it will most likely get worse since the draconian actions taken, ranging from lockdowns to school closures have not yet ended.

No one even knows the goal, if there even is a goal. We flattened the curve, the hospitals were empty, and cases are only rising because tens of thousands of people are being tested every day. Most of those being tested are asymptomatic and many are only being tested as a condition of employment or attendance at school. The conversation seems to have shifted to "safety."

But what does safety mean? Does it mean no cases? Does it mean that there are no deaths from COVID? If this is the case, we will never regain our freedom. There is no such thing as elimination of all risk. People are injured or die every year as a result of car accidents, flying in airplanes, and diving. People are struck by lightning, they fall while mountain climbing, and they can be injured while performing household tasks like mowing the lawn or painting while standing on a ladder.

Humans understand the risks associated with these activities and choose to do them anyway. Historically we have allowed people to make these decisions because we lived in a free society that allows people to take personal responsibility while making their own choices. This is not the case now, and if things do not change soon, we may not ever regain our freedom.

ENDNOTES

1. Provisional CoVID-19 Death Counts by Sex, Age... Centers for Disease Control. https://data.cdc.gov/NCHS/Provisional-COVID-19-Death-Counts-by-Sex-Age-and-S/9bhg-hcku/data accessed 8.8.2020

2. IBID

3. IBID

4. Holt-Lunstad J, Smith TB, Layton JB. "Social Relationship and Mortality Risk: A Meta-analytic Review." *PLoS Med* 2010 Jul;7(7):e1000316

5. Brooks SK, Webster RK, Smith LE et al. "The psychological impact of quarantine and how to reduce it: rapid review of the evidence." *Lancet* 2020 Mar;395(10227):912–920

6. Holt-Lunstad J, Smith TB, Baker M, Harris T, Stephenson D. "Loneliness and social isolation as risk factors for mortality: a meta-analytic review." *Perspect Psychol Sci* 2015 Mar;10(2):227-237

7. Social Isolation and Loneliness in Older Adults.: Opportunities for the Health Care System. The National Academies of Sciences, Engineering, Medicine. https://www.nationalacademies.org/our-work/the-health-and-medical-dimensions-of-social-isolation-and-loneliness-in-older-adults accessed 9.2.2020

8. Brooks SK, Webster RK, Smith LE et al. "The psychological impact of quarantine and how to reduce it: rapid review of the evidence." *The Lancet* 2020 Mar;395(10227):P912-920

9. Wu P, Fang Y, Guan Z et al. "The psychological impact of the SARS epidemic on hospital employees in China: exposure, risk perception, and altruistic acceptance of risk." *Can J Psychiatry* 2009 May;54(5):302-311

10. Loades ME, Chayburn E, Higson-Sweeney N et al. "Rapid Systematic Review: The Impact of Scial Isolation and Loneliness on the Mental Health of Children and Adolescents in the Context of COVID-19." *J Am Academ Child Adolesc Psychiatry* 2020 Jun;S0890-8567(20)30337-3

11. Sujata Gupta. "Social distancing comes with psychological fall-out." *ScienceNews* March 29, 2020 https://www.sciencenews.org/article/coronavirus-covid-19-social-distancing-psychological-fallout accessed 9.2.2020

12. Noah Higgins-Dunn, Kevin Breuninger. Cuomo says it's 'shocking' most new coronavirus hospitalizations are people who had been staying at home." *CNBC* May 7 2020 https://www.cnbc.com/2020/05/06/ny-gov-cuomo-says-its-shocking-most-new-coronavirus-hospitalizations-are-people-staying-home.html accessed 9.2.2020

13. Dave DM, Friedson AL, Matsuzawa K, Sabia JJ, Safford S. "Black Lives Matter Protests, Social Distncing and COVID-19." National Bureau of Economic Research June 2020 https://www.nber.org/papers/w27408.pdf accessed 9.2.2020

14. https://coronavirus.ohio.gov/wps/portal/gov/covid-19/dashboards/long-term-care-facilities/mortality accessed 8.8.2020

15. COVID-19 Dashboard. Ohio Department of Health. https://coronavirus.ohio.gov/wps/portal/gov/covid-19/dashboards/key-metrics/mortality accessed 8.8.2020

16. Thalia Brionez. Gatesville nursing home residents looking for pen pals during pandemic. *ABC25* August 4 2020 https://www.kxxv.com/news/local-news/gatesville-nursing-home-residents-looking-for-pen-pals-during-pandemic accessed 8.8.2020

17. Peter R. Breggin and Ginger Ross Breggin. CoVID-19 SOS: Saving America from Governor Cuomo, https://breggin.com/covid-19-sos-saving-america-from-gov-cuomo/ accessed 9.2.2020

18. Robert M. Sauer, Donald Siegel, David Waldman. CDC Has Become Centers for the Destruction of Childhood American Institute for Economic Research. June 25 2020 https://www.aier.org/article/cdc-has-become-centers-for-the-destruction-of-childhood/ accessed 9.2.2020

19. Dana Goldstein, Adam Popescu and Nikole Hannah-Jones. As School Moves Online, Many Students Stay Logged Out. *New York Times* April 8 2020 https://www.nytimes.com/2020/04/06/us/coronavirus-schools-attendance-absent.html accessed 9.2.2020

20. Dana Goldstein. Research Shows Students Falling Months Behind During Virus Disruption. *New York Times* June 5 2020 https://www.nytimes.com/2020/06/05/us/coronavirus-education-lost-learning.html accessed 9.2.2020

21. Fernando Alfonso III. The pandemic is causing and exponential rise in the exploitation of children, experts say. *CNN* May 25 2020 https://www.cnn.com/2020/05/25/us/child-abuse-online-coronavirus-pandemic-parents-investigations-trnd/index.html accessed 9.2.2020

22. Amy Hollyfield. Suicides on the rise amid stay-at-home order, Bay Area medical professionals say. *ABC7* May 21 2020 https://abc7news.com/suicide-covid-19-coronavirus-rates-during-pandemic-death-by/6201962/ accessed 8.8.2020

23. London RT. Is COVID-19 Leading to a Mental Illness Pandemic? *Medscape* April 3 2020

24. Paul Davidson. Unemployment soars to 14.7%, job loses reach 20.5 million in April as coronavirus pandemic spreads. *USA Today* May 8 2020 https://www.usatoday.com/story/money/2020/05/08/april-jobs-reports-20-5-m-become-unemployed-covid-19-spreads/3090664001/ accessed 9.2.2020

25. The Employment Situation – July 2020. Bureau of Labor Statistics. August 7 2020 https://www.bls.gov/news.release/pdf/empsit.pdf accessed 9.2.2020

26. Jim Sergent, Ledyard King, Michael Collins. 4 coronavirus stimulus packages $2.4 trillion in funding. See what that means to the national debt. https://www.usatoday.com/in-depth/news/2020/05/08/national-debt-how-much-could-coronavirus-cost-america/3051559001/ accessed 9.2.2020

27. Martin Hutchinson. The Coronavirus Economy Will Bring Inflation. April 23 2020. https://www.nationalreview.com/2020/04/the-coronavirus-economy-will-bring-inflation/ accessed 9.2.2020

28. Libby Kane and Tanza Loudeback. The IRS has sent over 159 million stimulus checks so far. Here's what to know if you are still waiting on yours. *Business Insider* Jun 23 2020 https://www.businessinsider.com/personal-finance/coronavirus-stimulus-check-questions-answers-2020-4 accessed 9.2.2020

29. Grace Segers. What's in the House Democrats' $3 trillion coronavirus relief bill? *CBS News* May 15 2020 https://www.cbsnews.com/news/coronavirus-relief-package-heroes-act-3-trillion-bill-house-democrats/ accessed 9.2.2020

30. Talia Kaplan. Art Laffer on market selloff amid coronavirus fears: 'Put your hands in your pockets' and wait it out. *Fox News* Mar 10 2020 https://www.foxnews.com/media/art-laffer-on-market-volatility-advise-investors-to-keep-hands-in-pockets-do-nothing accessed 9.2.2020

31. Emel Akan. Economists: World Slumps Into Worst Recession in Decades. *The Epoch Times* April 15 2020 https://www.theepochtimes.com/economists-world-slumps-into-worst-recession-in-decades_3313512.html accessed 9.2.2020

32. Brahma Chellaney. Opinion: the World Health Organizationmust stop covering up China's mistakes. *Project Syndicate* April 23 2020

https://www.marketwatch.com/story/the-who-has-a-big-china-problem-2020-04-22 accessed 9.2.2020

33. Joseph Bosco. The Wuhan virus and regime change in Washington. *The Hill* March 19 2020 https://thehill.com/opinion/national-security/488141-the-wuhan-virus-and-regime-change-in-washington accessed 9.2.2020

34. How China is Planning to use the coronavirus crisis to its advantage. *Washington Post* March 16 2020 https://www.washingtonpost.com/opinions/2020/03/16/how-china-is-planning-use-coronavirus-crisis-its-advantage/ accessed 9.2.2020

35. IBID

36. Coronavirus: World risks 'biblical' famines due to pandemic – UN." *BBC News* April 21 2020 https://www.bbc.com/news/world-52373888 accessed 9.2.2020

37. Karin Strohecker. Coronavirus crisis could plunge half a billion people into poverty: Oxfam. *Reuters* April 9 2020 https://www.reuters.com/article/us-health-coronavirus-poverty/coronavirus-crisis-could-plunge-half-a-billion-people-into-poverty-oxfam-idUSKCN21R0E7 accessed 9.2.2020

38. Apoorva Mandavilli. The Biggest Monster is Spreading. And It's Not the Coronavirus. *New York Times* Aug 3 2020 https://www.nytimes.com/2020/08/03/health/coronavirus-tuberculosis-aids-malaria.html accessed 9.2.2020

ENTER GEORGE FLOYD

AFTER MONTHS of being told to shelter at home, to keep businesses closed, to social distance, wear masks and follow the orders of government rulers "or else," life in the U.S. became even more complicated.

On May 25, 2020, an African American man, George Floyd, was arrested by the Minneapolis police after purchasing cigarettes with a counterfeit $20 bill. During a struggle with police, Floyd was pinned to the ground. Video footage showed that one officer kept his knee on Floyd's neck for over seven minutes. Floyd lost consciousness and died. All four officers were fired. One officer was later charged with second degree murder and the others were charged with aiding and abetting second degree murder.

Justifiable outrage ensued – and not just in the African American community. ALL Americans were outraged, as they should have been. Police brutality is inexcusable regardless of the perpetrator's race or the victim's race. Tens of thousands of people began protesting, mainly in major cities. This is an understandable response.

But what happened next was shocking. Protests turned violent. Windows were broken, buildings were burned, and businesses were looted. Tens of millions of dollars of property was destroyed.

Americans were already polarized over COVID-19, and divided into two camps, those who "believed" that there was a pandemic and that the lockdowns and restrictions were justified, and those who recognized that the data just did not add up. The divide deepened further due to the response to the protests and the destruction.

People had a right to peacefully protest in response to what happened to George Floyd, but many people also had a right to ask these questions: If it is ok to allow tens of thousands of people to gather close together, mostly without masks, and to protest for several days in a row, why is it not ok to allow 200 people to sit together in church? Why is it not ok to allow a restaurant to be fully occupied? Why are school officials talking about maintaining six feet between desks for little school children? Why, if gatherings were so dangerous, and masks were so necessary, were some of our rulers openly participating and showing up at these events without masks?

Here are some examples. The Empress Whitmer in Michigan imposed some of the worst lockdown orders in the country, going as far as forbidding people to engage in activities like planting gardens or using their boats, yet she was shown in several pictures standing shoulder to shoulder protesting with other government officials and demonstrators.[1] According to her office she did not violate her own executive order. But she did! Her order specifically stated "Persons may engage in expressive activities protected by the First Amendment within the State of Michigan but must adhere to social distancing measures recommended by the Centers for Disease Control and Prevention, including remaining at least six feet from people from outside the person's household."[2] You can view the pictures online – Whitmer is standing less than 6 inches from the persons around her - not maintaining 6 feet.

In the ultimate act of hypocrisy, the empress warned only 30 days before her appearance at the protests that if anti-lockdown protesters showed up at the Capitol, she would extend the lockdown orders even longer. In other words, if some Michigan residents disobeyed, all Michigan residents would suffer. After all, she is the Empress Whitmer and she can do anything she wants to her subjects.

The King that now rules New Jersey, Phil Murphy, said that protesting police brutality is much more important than protesting because your business is closed, you are going bankrupt, and you cannot feed your family. "I don't want to make light of this, and I'll probably get lit up by everyone who owns a nail salon in the state," Murphy said during a Monday briefing. "But it's one thing to protest what day nail salons are opening, and it's another to come out in peaceful protest, overwhelmingly, about somebody who was murdered right before our eyes."[3] What a tone-deaf thing to say! The person operating the nail salon and the employees who work for him/her cannot make a living while the business is closed. And King Phil does not think this matters? Of course not, as the lockdown has not hurt Murphy financially, the power-hungry despot could not care less about others and he's openly showing it.

Murphy went on to say, "The decision to go out or not go out, as long as you do it responsibly, safely and peacefully that's a decision, I would say, in this particular instance, I would leave to the individuals." Funny, almost all of Murphy's edicts from on high have not been left "to the individuals." Individuals still cannot open their businesses or go back to work. His hypocrisy is disgusting. What if his decision had been different and he had said, "We must allow businesses to re-open" and had disallowed protests? This would have been equally disgusting. Either way, his decisions were capricious, and we should all be appalled.

In Washington, D.C., Mayor Muriel Bowser visited a "Black Lives Matter" mural and removed her mask in order to pose with supporters and officials, all of them standing close together. It is always important to pose for pictures, and we all know that the virus does not pass from person to person during picture taking. Right. [4]

Public health experts who were enthusiastic supporters of lockdowns also changed their minds about the need to remain at home and social distance when the protests started. "We should always evaluate the risks and benefits of efforts to control the virus," tweeted Jennifer Nuzzo, a Johns Hopkins epidemiologist. "In this moment the public health risks of not protesting to demand an end to systemic racism greatly exceed the harms of the virus."[5]

This sounds exactly like statements made about the lockdown, that the health risks associated with it are exceeding the benefits. An example is the doctors in San Francisco who reported that more people were dying of suicide than COVID in their area and perhaps it was time to end the lockdown. Yet the lockdowns in California continue, while the protests in California are massive at the same time. We think it is a fair question to ask why the benefits of protesting exceed the risks of harm from the virus, but the benefits of preventing suicides, overdose, deaths, bankruptcies, homelessness and food insecurity are treated differently.

Abraar Karan, a physician at Brigham and Women's Hospital, was an avid supporter of lockdowns, another member of the ruling class that did not have to worry about money or meeting a payroll or feeding his children. He showed no compassion for other people who were negatively impacted by lockdowns. However, he tweeted this about the protests: "The injustice that's evident to everyone right now needs to be addressed. While I have voiced concerns that protests risk creating more outbreaks, the status quo wasn't going to stop #covid19 either."[6] In other words, he has suddenly changed his mind in two ways. He thinks that staying at home does not stop the spread of COVID-19, and he has found a reason HE THINKS justifies going outside. Of course, everyone should have their reasons for wanting to leave the house or go to work vetted by a member of the ruling class like Karan, who knows so much more than the peasants who are clearly beneath him.

Tom Frieden, former director of the CDC, had just recently cautioned against opening the economy too fast. But he changed his mind and supported mass protests. He and others claimed that if we did not address racial inequality now, it will be harder to fight COVID-19.[7] We are trying to figure out what this means – the virus will become more contagious or lethal if it does not approve of how we handle racial inequality?

For the record, we think it is important to address racism. We despise it, don't practice it, one of us has been subjected to discrimination for other reasons, and we both would like to end it once and for all. But we can still think that what these people are saying makes no sense at all, and that the inconsistencies deserve some attention.

Hundreds of public health workers signed an open letter in response to these events that was almost incomprehensible, stating that demonstrations against the lockdown were rooted in white nationalism and represented disrespect for Black lives.[8] Apparently the need to earn a living, educate your kids, and feed your family are now considered a demonstration of white nationalism. This is ridiculous and it ignores the fact that a person can hold two thoughts at the same time. A white person can be against the senseless killing of black people AND still wonder why he cannot go back to work or visit a library.

The letter states, "Protests against systemic racism, which fosters the disproportionate burden of COVID-19 on Black communities and also perpetuates police violence, must be supported. Staying at home, social distancing, and public masking are effective at minimizing the spread of COVID-19. However, as public health advocates, we do not condemn these gatherings as risky for COVID-19 transmission."[9] What? Staying at home is protective, but they don't see going out and gathering as risky. Which is it? In this case it can't be both because people cannot stay home and go out and protest at the same time.

Perhaps the most hypocritically transparent response came from Bill de Blasio, the ruler of New York City. Throughout the lockdown he specifically called out Jewish communities who disobeyed him by having large funerals. When a reporter asked him why he was continuing to allow masses of people to be out and about protesting when religious services were forbidden, he replied, "Four hundred years of American racism. I'm sorry, that is not the same question as the understandably aggrieved store owner or the devout religious person who wants to go back to services."[10]

There is either a public health threat or there is not. De Blasio's response is scientifically illiterate. A lethal virus cannot possibly care about why large numbers of people are gathering. If it is dangerous to attend a funeral, it must be equally dangerous to be in a crowd of thousands protesting.

Still others argue that the virus does not spread easily outdoors, and the protests are outdoors which makes protesting safe. Again, this makes no sense. Beaches and playgrounds were closed and many still are. This must be another instance in which the virus is very choosy and rewards some behaviors while punishing others. The virus only likes for people to be outside for a specific purpose. Laying on the beach or playing on a playground – not good. The virus does not like this, spreads like crazy and becomes more virulent. Protesting — the virus likes this and stands down.

There are several points we want to make in reference to these most unfortunate events.

First, the idiots making these statements are trying to do what our rulers have been trying to do – turn us against one another. We must not let this happen. We can be outraged at the needless death of a black American AND at the government's irresponsible decisions at the same time. **We should all be together on all of this because we are all affected by all of this.**

Second, let's consider for a minute that no one in the media is asking any of our rulers questions such as how they can justify the collateral damage to millions of people due to lockdowns, restrictions and closures, and then tell everyone it's ok to protest in groups of thousands or even tens of thousands? Most members of the media have become criminal co-conspirators. They have contributed to the death and destruction that has resulted from the irresponsible and reckless decisions that have been made by health and government officials. One of the jobs of the media is to hold government officials responsible and accountable. One of the reasons why our rulers have been able to get away with what they have done is because reporters have not done their jobs.

Columnist Bethany Mandel, who called for the lockdowns to end and was called "the Grandma Killer" in a trending hashtag, says, "These people have killed tens of thousands of seniors in nursing homes. They won't let us go out of our own homes for months. They've destroyed the economy. And now if you don't do what they've been telling us not to do all this time - to gather with other people - we're irredeemable racists. This is the scandal of the century. We've all been played."[11]

Yes, we have been played. The smart thing would be for the people who did this to come clean and admit they were wrong.

We could stop the nonsense and start rebuilding our country again. We the people would be a lot more forgiving if they were to do this. But they show no sign of doing this, so we the people are going to have to fix this ourselves. We cannot fix it if we let them wreak even more havoc by turning us against one another. This awful situation has affected everyone negatively, regardless of race, gender, religion, or any other identifying characteristic.

Martin Niemoller was a prominent Lutheran pastor in Germany. He emerged as an outspoken public foe of Adolf Hitler and spent the last seven years of Nazi rule in concentration camps. He is perhaps best remembered for his postwar words:

> "First they came for the socialists, and I did not speak out—because I was not a socialist.
>
> Then they came for the trade unionists, and I did not speak out— because I was not a trade unionist.
>
> Then they came for the Jews, and I did not speak out— because I was not a Jew.
>
> Then they came for me—and there was no one left to speak for me"

We need to stand up for each other while we still can. We are better than this. Our hearts are big enough to want equality for everyone in every area. All of us deserve to be safe from law enforcement. All of us are entitled to the freedoms outlined in our constitution. And all of us are going to have to work together to protect ourselves and restore our rights.

ENDNOTES

1. Adam Shaw. Officials who pushed strict lockdowns now argue protesters are an exception. *Fox News* June 5 2020 https://www.foxnews.com/politics/officials-lockdowns-protesters-exception accessed 9.2.2020

2. Executive Order 2020-110 FAQs https://www.michigan.gov/coronavirus/0,9753,7-406-98178_98455-530654--,00.html accessed 9.2.2020

3. Adam Shaw. "Officials who pushed strict lockdowns now argue protesters are an exception." *Fox News* June 5 2020 https://www.foxnews.com/politics/officials-lockdowns-protesters-exception accessed 9.2.2020

4. IBID

5. IBID

6. Dan Diamond. "Suddenly, Public Health Officials Say Social Justice Matters More Than Social Distance." *Politico* Jun 4 2020 https://www.politico.com/news/magazine/2020/06/04/public-health-protests-301534 accessed 9.2.2020

7. IBID

8. Open letter advocating for an anti-racist public health response to demonstrations against systemic injustice occurring during the COVID-19 pandemic. https://drive.google.com/file/d/1Jyfn4Wd2i-6bRi12ePghMHtX3ys1b7K1A/view accessed 9.2.2020

9. IBID

10. Mark Hemingway. The Media's Double Standard on Protest Coverage. *Daily Signal* June 7 2020 https://www.dailysignal.com/2020/06/07/the-medias-double-standard-on-protest-coverage/ accessed 9.2.2020

11. https://twitter.com/bethanyshondark/status/1268539566090268672 accessed 9.2.2020

WHICH NEW NORMAL WILL WE CHOOSE?

THE COVID-19 debacle has been described as a dystopian nightmare from which we cannot seem to awaken. If the people who planned this event get their way, the science fiction-type existence we have experienced for a few months will become permanent. It is referred to by "experts" and the media as "the new normal."

There has been no attempt to hide what the drug companies, Bill Gates, and the global elites have in mind for the future. It is living life in a perpetual state of emergency that justifies several rules including mandatory vaccines, immunity passports, continued restrictions on personal freedoms, more government control, constant surveillance, social distancing, and a rebuilding of the world based on unelected people who have put themselves in charge.

The World Economic Forum (WEF) and The Great Reset

The World Economic Forum was founded by Professor Klaus Schwab and was originally called the European Management Forum. WEF is a non-profit organization based in Geneva and its original goal was to help European companies to adopt U.S. management practices, and to promote good corporate citizenship. The organization advised that companies should consider clients, customers, employees, and the communities in which they were based when making their decisions.

In 1987 the organization was renamed the World Economic Forum (WEF) and its stated mission changed. WEF decided to be the global platform for "public-private cooperation."[1]

Founder Schwab has some interesting ideas. In an essay posted on the WEF site, Schwab endorses stakeholder capitalism, and proposes that private corporations should be trustees of society. He thanks Greta Thunberg for "reminding us that adherence to the current economic system represents a betrayal of future generations" due to its lack of environmental sustainability. You just have to love a powerful guy who gets his advice from a depressed teenaged high school dropout![2]

WEF's strategic partners include the **Bill and Melinda Gates Foundation**, Facebook, Google, Microsoft, Wellcome Trust and several drug companies.[3] Many important announcements concerning vaccines are made at annual meetings of the WEF in Davos, including the launch of GAVI by the Bill and Melinda Gates Foundation. Vaccines for COVID-19 are an important part of "the new normal" envisioned by WEF and its allies.

Within a short period of time after the pandemic was declared, the WEF launched an elaborate COVID-19 platform with thousands of pages of content. The site is titled "The Great Reset."[4] We encourage you to spend some time exploring this site. Even months later it is difficult to believe that this amount of information was assembled after the pandemic was declared. A date was already posted for January 2021 for an event centered around The Great Reset theme.[5]

In documents posted on the site, Schwab seems almost delighted about the pandemic, writing "The pandemic represents a rare but narrow window of opportunity to reflect, reimagine, and reset our world." He urges a "Great Reset" which would "steer the market toward fairer outcomes," and encourages governments to enact reforms that include changes to wealth taxes, withdrawal of fossil fuel subsidies, and new rules governing intellectual property, trade, and competition. The Great Reset agenda should ensure that investments focus on equality and sustainability, green infrastructure, and incentives for companies to improve their environmental, social and governance metrics.[6] The announcement of the "Great Reset" was made by the Prince of Wales and Schwab at a virtual meeting in June, only a few weeks after the lockdowns had occurred.

Schwab's comments are eerily reminiscent of those made by Former WHO Director-General Margaret Chan's following the fake H1N1 pandemic which she helped to engineer a few years ago. She said in a speech that "ministers of health" should take advantage of the "devastating impact" swine flu will have on poorer nations to get out the message that "changes in the functioning of the global economy" are needed to "distribute wealth on the basis of" values "like community, solidarity, equity and social justice." She further declared that the pandemic should be used as a weapon against "international policies and systems that govern financial markets, economies, commerce, trade, and foreign affairs."[7] In other words, Chan looked at fake pandemics as a form of social engineering, to be executed according to her beliefs, of course.

Apparently, pandemics give smart elite people the opportunity to rework the world as they see fit. The WEF platform clearly states that the "old system" must be replaced by a "… new social contract" that ensures that "…we live up to the expectations of young people." Referring to a "Fourth Industrial Revolution," the WEF proposes that the 2021 meeting will bring together "…key government and business leaders in Davos…" The global media and social media, the very organizations that controlled the COVID-19 narrative, are mentioned as vehicles to mobilize millions of people to join in the remaking of the world.

An important question to ask is whether or not the world's ordinary citizens – people like us and like you – want the Prince of Wales, Greta Thunberg, Facebook, Google, and Schwab deciding how the world will be organized and how we will live from this point forward.

Other Global Organizations Mobilized Too

The Global Fund, along with board member Dr. Deborah Birx, sprang into immediate action, committing up to one billion dollars to help countries fight COVID-19. The money was allocated to "accelerate the development, production, and equitable global access to safe, quality, effective and affordable COVID-19 **diagnostics, therapeutics, and vaccines…**"[8] These three words became literally a mantra, which were repeated again and again by drug companies, world leaders, and health authorities.

The European Union and its partners hosted an international pledging conference starting Monday, May 4, 2020[9] at which the Bill and Melinda Gates Foundation pledged $100 million dollars. The event was hosted by Yrsula von der Leyen, President of the European Union, and one country leader after another repeated the mantra, along with his/her pledge, that money was needed for **diagnostics, treatment, and safe vaccines.**[10]

Almost $7.4€ was raised, and the representatives of the organizations that would receive the money, such as WHO, CEPI and GAVI, all gave testimonials about the importance of **diagnostics, therapeutics, and vaccines.**

Paul Farmer and his organization, Partners in Health, mobilized immediately to contract with state and local governments for "contact tracing" services. Some "experts," such as Tom Frieden, former Director of the CDC, estimated that as many as 300,000 people will need to be hired to fill demand.[11]

An announcement was made by **Rockefeller Foundation** President Rajiv Patel on April 20, 2020 that the Rockefeller Foundation was investing $100,000,000 to perform COVID-19 testing and contact tracing. The plan was to begin with three million tests per week, ramping up to 30 million per week in 6 months. There are approximately 330,000,000 million people in the U.S., which means that the foundation's goal is to test all of us. As many as 300,000 people will be hired as part of what Patel calls a "community health corps" to execute the plan. Those who have been exposed to COVID-19 will be identified and quarantined. According to Patel, this is the only way that Americans can safely go back to school and to work.[12]

The Real Goal: Mandatory COVID-19 Vaccines

Bill Gates has stated clearly that his goal is universal vaccination. He wrote in GatesNotes on April 30, 2020, "One of the questions I get asked the most these days is when the world will be able to go back to the way things were in December before the coronavirus pandemic. My answer is always the same: when we have an almost perfect drug to treat COVID-19, or when almost every person on the planet has been vaccinated against coronavirus."

Gates states that a treatment is unlikely to be discovered anytime soon, ignoring considerable evidence that hydroxychloroquine is both inexpensive and effective. He says the world cannot function normally until everyone (over 7 billion people) receives a vaccine. In essence Gates is declaring that he will decide the fate of the world and its citizens.[13]

Fauci agrees that the world will not return to normal until a vaccine is available.[14] In late March/early April, he said that a COVID-19 vaccine would not be ready for at least 12-18 months, and that such a vaccine would be needed before many restrictions could be lifted. Many vaccine advocates, such as Dr. Paul Offit, said "When Dr. Fauci said 12-18 months, I thought that was ridiculously optimistic, and I'm sure he did too."[15] It's almost unthinkable that the lives of seven billion people could be placed on hold for this length of time, but in COVIDLAND, almost anything seems possible.

Like so many statements Fauci has made, his predictions about a COVID-19 vaccine changed. Just a few weeks later he said that hundreds of millions of doses could be made available by early 2021. His confidence lies, in part, because he says he's part of the team that is involved in developing the vaccine.[16] Gates stated that based on this timeline, safety and efficacy might be compromised, and also that the vaccine may only be effective for a few months, which means repeated vaccinations will be necessary. What a windfall for drug companies! To return to normal 7 billion people must agree to receive not just one, but several COVID-19 vaccines.[17]

The problem is that it is highly unlikely that an effective vaccine for COVID-19 will ever be developed.

A Short History of Flu Vaccines

Today, the American Academy of Pediatrics, the Advisory Committee on Immunization Practices, the Centers for Disease Control, and the World Health Organization all aggressively promote flu vaccines. The origin of this recommendation goes all the way back to the 1918-1919 flu pandemic, which killed about 50 million people worldwide. There was little understanding of how the epidemic occurred, but doctors started promoting vaccines to prevent influenza. Early flu vaccines were tested in the military, but by 1947 it was determined that "the incidence of disease was no different in vaccinated and unvaccinated individuals."[18] In spite of this, the vaccine was promoted for use in the general population.

In early 1957 an outbreak of the Asian flu began in China. Concerned about another epidemic, Maurice Hillman at Walter Reed Army Hospital sent virus samples to drug makers and encouraged them to make a vaccine. The epidemic eventually caused almost two million deaths worldwide, 70,000 of those deaths in the U.S.[19] Millions of doses of vaccine were given to Americans, but the vaccine was again proven to be worthless.[20]

Vaccine advocates claimed that the historic failure of vaccines was because they were given too late and put forth the theory that starting before a flu outbreak would result in a higher efficacy rate. In response, in 1960 public health officials started recommending routine vaccination, which became a public policy within a few years with virtually no additional data to support such a policy. In fact, the evidence pointed to routine flu vaccines as a major public policy failure. CDC chief epidemiologist Alexander Langmuir and colleagues wrote in a 1964 paper that they "...reluctantly concluded that there is little progress to be reported. The severity of the epidemic of 1962-63 ...demonstrates the failure to achieve effective control of excess mortality."[21] They went on to say that routine vaccination should be continued only if better evidence could be found to justify the significant cost of the vaccination program.

The CDC conducted a randomized, double-blind trial designed to determine if the flu vaccine prevented morbidity and mortality, and concluded, "Despite extensive use of influenza vaccines...attainment of (improved morbidity and mortality) has never been demonstrated."[22] A Food and Drug Administration review arrived at the same conclusion, and cautioned that there were methodological flaws in many of the studies reviewed.[23]

Today's Recommendations

Based on the history of the vaccine, it is not surprising that continuing to market the flu vaccine to the public requires considerable misrepresentation. This starts with overstating both the incidence of and risks associated with getting the flu. First, we are all exposed to flu viruses all the time; the flu virus is constantly present and does not make a brief appearance during "flu season." Another issue is that influenza is often confused with influenza-like illness (ILI) which can result from 200 viruses in addition to influenza A and B. These viruses produce the same symptoms as flu, which include fever, headache, aches, pains, cough, and runny noses, making it impossible to distinguish between the two without diagnostic testing. An individual is seven times more likely to have an influenza-like illness than influenza, but ILI is rarely serious.

Nonetheless the Centers for Disease Control promotes flu vaccines, stating, "Influenza is a serious disease that can lead to hospitalization and sometimes even death. Every flu season is different, and influenza infection can affect people differently. Even healthy people can get very sick from the flu and spread it to others."[24]

Although the CDC offers conflicting narratives about the flu; on another page of its website, the agency states, "CDC does not know exactly how many people die from seasonal flu each year."[25] In other words, the CDC aggressively promotes a solution for a problem that it cannot quantify.

Safety of Flu Vaccines

What can be more easily quantified is risks associated with the vaccine. On several occasions, flu vaccine programs have been terminated due to side effects. In October 1976, The National Influenza Immunization Program (NIIP) started with about one million vaccinations per week for a new strain of swine flu, and this number grew quickly to four million per week. Within only two months, ten states had reported cases of Guillain-Barre syndrome (GBS) linked to the vaccine. In December 1976, the program was discontinued.

By January 1977, more than 500 cases of GBS had been reported. Some patients recovered completely, some partially, and 25 people died. The NIIP determined that the risk of developing GBS within 6 weeks was 10 times higher for those receiving a flu vaccine than for unvaccinated people. While this should have been the end of promoting population-wide vaccination for flu until safety could be established, flu vaccine promotion programs continued.

This debacle took place at a time when news reporters aggressively investigated government agencies and health professionals and reported their misbehavior. CBS's hit show *60 Minutes* featured a lengthy story about the flu vaccine. Wallace announced at the beginning of the segment that Washington D.C. had decided that all Americans should get a flu vaccine to prevent what the government predicted would be a pandemic. Forty-six million Americans received a vaccine, and eventually almost 4000 Americans filed lawsuits against the U.S. government for claims totaling $3.5 billion dollars. Two thirds of these claims were for neurological damage or death.

According to Wallace's report, the swine flu "pandemic" started at Fort Dix in New Jersey in January 1976. One soldier who felt sick participated in a forced march and later died. Four others became sick and recovered. An army doctor sent samples of throat cultures from several soldiers to the New Jersey Department of Health for testing. The department reported that many samples were the common flu, but they could not identify the virus that sickened the four soldiers who recovered or the one who died. These samples were forwarded to the Centers for Disease Control and Prevention, where they were identified as the swine flu. Remember, there were a total of five affected individuals in the U.S. at this time.

Dr. David Sencer, who was head of the CDC at the time, oversaw the rapid development of a vaccine for the swine flu and an aggressive program to promote it. Wallace asked how many cases had been confirmed around the world at the time that the program was launched and he replied that there were several reported but none were confirmed, and acknowledged that there were no outbreaks of swine flu confirmed anywhere in the world.

The CDC's consent form stated that the vaccine had been tested but did not disclose the one being administered was different from the tested vaccine. The form included nothing about the risks of many of the side effects experienced by recipients including GBS and death.

Wallace asked Sencer if he knew that there were risks of neurological damage due to the vaccine, and Sencer replied that he had no idea. Dr. Michael Hattwick, who directed the surveillance team for the swine flu program at the CDC, told a different story. He says he knew of cases of GBS and told his superiors, and that Sencer did know.

The CDC launched an aggressive promotional campaign that included a list of famous people who had already received the vaccine. These included then-President Gerald Ford, Elton John, Mary Tyler Moore, and Ralph Nader. When asked about the vaccine on camera, Mary Tyler Moore told Wallace that she did not receive the vaccine and that she had not given the CDC permission to use her name in its campaign.[26]

This was not the end of flu vaccines, however. Aggressive campaigns were launched in 1992, 1993, and 1994, and were again were shown to increase the risk of GBS.[27] [28] [29]

As of November 2013, there were 93,000 reactions attributed to flu vaccines reported to the Vaccine Adverse Event Reporting System (VAERS) including 1080 deaths, 8888 hospitalizations, 1801 disabilities, and 1700 cases of Guillain Barre Syndrome.[30]

Data from the National Vaccine Injury Compensation Program shows that the flu shot is the most dangerous vaccine in America. During one reporting period, out of 134 cases settled before the court, 79 were due to the flu shot, including three deaths. Guillain-Barre syndrome was the most common injury, but others included acute disseminated encephalomyelitis, transverse myelitis, shingles (herpes zoster), neuropathic demyelination, seizures, neuropathy, brachial plexopathy, rheumatoid arthritis, optic neuritis, and Bell's palsy.[31]

The adjuvants in flu vaccines include mercury (25 mcg), formaldehyde, polyethylene glycol, egg protein, polysorbate 80, MSG, pig gelatin, and antibiotics. Between 2009 and 2010, fetal deaths reported to VAERS had increased significantly due to the addition of thimerosal.[32]

Continuing Issues with Efficacy

The side effects are concerning and become even more so when considering the vaccine's continued poor performance. A 2005 study concluded that the benefits of the flu vaccine were overstated, and "...[even during two pandemic seasons] the estimated influenza-related mortality was probably very close to what would have occurred had no vaccine been available."[33]

Cochrane Collaboration is the most independent medical research organization in the world. A Cochrane review analyzed the impact of flu vaccines on healthy adults, including pregnant women and newborns, by looking at 90 reports of 116 studies that compared flu vaccines to placebo or to no intervention. Combined, the studies included close to ten million people. Cochrane concluded that 40 people would have to be vaccinated to prevent just one case of influenza-like illness (ILI), and 71 people would have to be vaccinated to prevent one case of influenza. The vaccine had no effect on number of working days lost or hospitalization rates. The vaccine also had almost no effect on pregnant women or their newborn babies. Live aerosol vaccine was similarly useless.[34] In another review, Cochrane reported that flu vaccines were not effective for the elderly either.[35]

Cochrane conducted a similar review to evaluate the efficacy rates (defined as prevention of confirmed influenza and influenza-like illness), and adverse events of influenza vaccines in healthy children. The review included 75 studies and showed:

- Six children under age six would have to be vaccinated with live attenuated vaccine to prevent one case of flu.
- In all the studies, there was no useable data for children under the age of two.
- For children age two or younger, inactivated flu vaccines were no more effective than placebo.
- To prevent one case of influenza in children over the age of six, twenty-eight children need to be vaccinated, and eight need to be vaccinated to prevent just one case of influenza-like illness.

The researchers found "no evidence of effect on secondary cases, lower respiratory tract disease, drug prescriptions, otitis media... (only) weak single study evidence of effect on school absenteeism and keeping parents from work." In other words, the children had almost no reduction in risk of developing the flu, flu-like illness, or of developing complications from flu. The vaccine was shown to be almost worthless.

Side effects were noted, however, and some were serious such as narcolepsy and febrile convulsions.

The researchers expressed surprise that the current recommendation is to vaccinate healthy children starting at 6 months of age in the U.S. and several other countries based on such limited evidence, and advised that research is needed in order to identify all potential harm resulting from flu vaccines.

Just as important, researchers identified issues concerning study design, funding, and scientific misbehavior. The Cochrane group reported that industry-funded studies showed more positive results than those funded with public money. They reported that "An earlier systematic review of 274 influenza vaccine studies published up to 2007 found industry-funded studies were published in more prestigious journals and cited more than other studies independently from methodological quality and size. The review showed that reliable evidence on influenza vaccines is thin but there is evidence of widespread manipulation of conclusions and spurious notoriety of the studies. The content and conclusions of this review should be interpreted in light of this finding."[36] According to Tom Jefferson, head of the Vaccine Field Group at the Cochrane Database Collaboration, "The vast majority of the studies (are) deeply flawed. *Rubbish* is not a scientific term, but I think it's the term that applies."[37]

Translation: A lot of scientific misconduct is required to report conclusions that support flu vaccines.

Even the package inserts on the vaccines state that they are not effective. For example, the package insert for FLULAVAL

2013-2014 formula for Influenza subtype A viruses and type B virus states, "...there have been no controlled trials adequately demonstrating a decrease in influenza disease after vaccinations with FLULAVAL."[38]

Flu Vaccines and Healthcare Workers

Despite this information, healthcare workers (HCW) are routinely forced to get an annual flu shot, and often threatened with termination if they refuse. One of the reasons is that reimbursement rates from Medicare/Medicaid are tied to vaccination rates for hospital staff. Hospital systems must have a 90% or higher vaccination rate or they lose 2% of their funding from these programs.[39]

The flu shot does not protect patients, since patients do not get the flu from asymptomatic healthcare workers, whether or not they have been vaccinated. A meta-analysis conducted by CDC researchers confirmed this, showing that flu vaccines for healthcare workers offer little protection. The analysis looked at four studies from long-term facilities or hospitals and concluded that the impact on lab-confirmed flu was not statistically significant. The researchers noted that there are no estimates available on the number of deaths from flu in frail elderly people. Furthermore, the researchers ranked the quality of evidence for HCW vaccination on mortality as moderate and the quality for both influenza and hospitalization as low.[40]

Physician Daniel O'Roark, an outspoken critic of mandatory vaccines, refers to flu season as the yearly "influenza hysteria and the absurdity known as mandatory vaccination of HCW." O'Roark states that it has until recently been considered absurd to mandate medical treatments of any type for people who are mentally competent; for minors and those who were incompetent, consent would be given by parents or those with legal power of attorney. The reason, according to O'Roark, is that all medical treatments, including vaccines, subject people to varying degrees of risk.[41]

Healthcare workers are fighting back. A New Jersey appeals court ruled in favor of a nurse after she was fired for refusing a flu shot without claiming a religious or medical exemption, stating that the hospital that employed her "unconstitutionally discriminated against" June Valent when she was unfairly denied unemployment benefits.[42]

Nationally, 17% of hospital workers refuse the flu vaccine, and during the 2014-2015 flu season, 30% of hospital workers in New Jersey, Florida, and Alaska refused. There are 3662 hospitals in the U.S. and 966 report that 25% of their workers say "no," and 140 report that half or more are saying "no" to flu vaccines.[43]

Flu Vaccines and Pregnant Women

Pregnant women are also coerced into getting flu vaccines. According to the CDC's website, "if you are pregnant, a flu shot is your best protection against serious illness from the flu. A flu shot can protect pregnant women, their unborn babies and even the baby after birth."[44] But the package insert for the H1N1 vaccine states, "It is not known whether these vaccines can cause fetal harm when administered to pregnant women or can affect reproduction capacity."

Another study concluded that flu vaccines cause an inflammatory response in pregnant women and that inflammation increases the risk of both preeclampsia and premature birth. The researchers added that more research is needed to determine that flu vaccines are safe.[45] The package insert for FLULAVAL states, "Safety and effectiveness of FLULAVAL have not been established in pregnant women or nursing mothers."[46]

The FDA states that unless vaccines are specifically intended to be used in pregnant women, pregnant women are not eligible to participate in clinical trials, and that if a woman becomes pregnant during a clinical trial, she should not receive any more vaccines.[47] Yet the American College of Obstetrics and Gynecology says, "Any of the licensed, recommended, age-appropriate, inactivated influenza vaccines can be given safely during any trimester. Therefore, it is critically important that obstetrician–gynecologists and other obstetric care providers recommend and advocate for the influenza vaccine."[48]

What We Can Expect From the COVID-19 Vaccine

Obviously, nothing has been learned from previous flu vaccine programs, and nothing has changed. The rush to develop a COVID-19 vaccine is on, and speed seems to be most important – much more than safety. On April 6 Bill Gates announced that his foundation would invest billions of dollars to fund companies that develop COVID-19 vaccines. "Our foundation is trying to be as helpful in a very constructive way as possible. And that's why

I've talked to the head of the pharmaceutical companies. We've talked to a lot of the agencies, including -- CDC and NIH about how we work together on the vaccines and the drugs."[49]

This time, risks associated with the vaccine might be even greater. Fauci acknowledged this in an interview with Howard Bauchner MD, Editor in Chief for *JAMA*, April 8, 2020. "One of the things you have to be careful of when you're dealing with a coronavirus is the possibility of enhancement." He said that some vaccines cause antibody dependent enhancement (AED), which paradoxically leaves the body *more* vulnerable to severe illness after inoculation.[50]

Other scientists are urging caution. Vaccines developed for feline infectious peritonitis virus increased risk of cats becoming infected, and the same result has been found in animal studies for other coronavirus vaccines.[51]

SARS-CoV vaccines induced antibody response but caused immunopathologic-type lung disease and excessive and dangerous host immune response and severe pneumonia when tested in mice.[52] [53] Researchers who conducted another study wrote, "These data raise significant concerns regarding [double-inactivated vaccine] DIV vaccine safety and highlight the need for additional studies of the molecular mechanisms governing DIV-induced eosinophilia and vaccine failure, especially in the more vulnerable aged-animal models of human disease."[54]

There are also considerable conflicts of interest concerning Moderna, a U.S. biotech company, which is the leading contender for bringing a COVID-19 vaccine to market. Its vaccine is the first one to be used in clinical trials. The vaccine uses a new technology called messenger RNA (mRNA).

Vaccines normally work by training the body to recognize and respond to viruses and bacteria. They are made from small or inactivated samples of a virus that result in an immune response when injected into the body.

mRNA vaccines trick the body into producing the viral proteins by using messenger RNA. When injected into the body, cells read the "instructions" to build a viral protein and create viral molecules. The immune system then produces antibodies to attack them.

This technology has never been approved for use before. There are many unknowns, including whether this technology is safe, and how long immunity would last. And then there are those recurrent conflicts of interest.

Fauci is head of the National Institute of Allergy and Infectious Disease. Scientists in his agency, in conjunction with researchers working for Moderna, developed mRNA technology[55] in collaboration with The Coalition for Epidemic Preparedness Innovation (CEPI)[56] and patented it.[57] Six of the researchers listed on the patent application work for the NIAID and are:

- Barney Graham, MD; Deputy Director NIAID Vaccine Research Center
- Kizzmekia S. Corbett, PhD: Scientific Lead NIAID's Coronavirus Vaccine Program
- M. Gordon Joyce, PhD: NIAID Vital Pathogenesis Laboratory

- Hadi M. Yassine, NIAID
- Masaru Kanekiyo PhD: NIAID Pathogenesis Translational Science Core
- Olubkola Abiona: NIAID Vital Pathogenesis Laboratory

Moderna's January 6, 2020 letter to shareholders included this statement:

"We believe mRNA vaccines have the potential to provide some critical advantages for preventing the spread of debilitating and deadly infections. A November 2019 review of novel vaccine technologies published in *Nature Reviews Immunology* and written by John R. Mascola (from NIH) and Anthony S. Fauci noted the potential of mRNA, stating, "The combination of preclinical and clinical data bodes well for the potential of mRNA vaccines to serve as a rapid and flexible platform that will be useful in responding to both seasonal and pandemic influenza, and by extension to any newly emerging infectious agent."[58] There is nothing quite like having the head of the NIAID on your side when trying to rush a new technology to market!

In February 2020, Moderna reported that it had shipped the first batch of the mRNA-1273 vaccine against novel coronavirus to the NIH's National Institute of Allergy and Infectious Diseases (NIAID) for use in a planned Phase I study in the U.S. The study is called "Safety and Immunogenicity Study of 2019-nCov Vaccine (mRNA-1273) to Treat Novel Coronavirus."[59] In the rush to develop a vaccine, researchers are being permitted to bypass the usual protocol that includes testing on animals and have moved directly to human trials. "Outbreaks and national emergencies often create pressure to suspend rights, standards and/or normal rules of ethical conduct. Often our decision to do so seems unwise in retrospect," wrote Jonathan Kimmelman.[60]

Bill Gates firmly believes in the technology, writing, "That's why I'm particularly excited by two new approaches that some of the candidates are taking: RNA and DNA vaccines. If one of these new approaches pans out, we will likely be able to get vaccines out to the whole world much faster."

"Our foundation—both through our own funding and through CEPI—has been supporting the development of an RNA vaccine platform for nearly a decade. We were planning to use it to make vaccines for diseases that affect the poor like malaria, but now it's looking like one of the most promising options for COVID. The first candidate to start human trials was an RNA vaccine created by a company called Moderna."

He added that "In order to stop the pandemic, we need to make the vaccine available to almost every person on the planet." He states that the Gates Foundation is working with CEPI, WHO, and governments to figure out how to make seven billion doses of vaccine; 14 billion doses if it turns out to be a multi-dose product.[61]

What a bonanza for drug companies. If Gates' and Fauci's dreams come true, and seven billion doses of COVID-19 vaccine are made – just one for almost every man, woman and child on the planet – at a cost of even $50 per dose the drug companies would take in $350 billion dollars *just for one dose!*

But there is more...

Bill Gates not only wants to see the entire world vaccinated, he has invested a considerable amount of money in technology that would allow vaccines to be delivered using a digital identity that keeps track of who has received vaccinations. The Bill and Melinda Gates Foundation has invested $21 million to develop vaccine technology that injects invisible nanoparticles under the skin that can be read by smartphones.

Kevin McHugh, assistant professor of bioengineering at Rice University describes the technology as "something like a barcode tattoo." "The Bill and Melinda Gates Foundation came to us and said, 'Hey, we have a real problem – knowing who's vaccinated. So our idea was to put the record on the person. This way, later on, people can scan over the area to see what vaccines have been administered and give only the ones still needed.'"[62]

Researchers have already published a study showing that the technology works with a vaccine against COVID-19.[63] Gates funded another study in which researchers from the Massachusetts Institute of Technology, the Institute of Chemistry of the Chinese Academy of Sciences in Beijing and the Global Good, Intellectual Ventures Laboratory in Bellevue, Washington determined that "near-infrared quantum dots" can be implanted under the skin along with a vaccine to encode information for "decentralized data storage and bio-sensing." The researchers created a platform capable of encoding data on individuals for at least 5 years after administration, and which could be read using a smartphone.[64]

Fauci agrees with this also, and in an interview on CNN on April 10, 2020 he said that the government is considering issuing certificates of immunity to Americans to identify who has been infected. He further said that he could imagine a time when people would be required to carry such identification in order to move around freely.[65]

This is not as futuristic as it might sound. A digital identity program, launched in 2018, is being implemented in poor, rural communities in West Africa right now. Trust Stamp is an "identity authentication" company that has partnered with the Bill Gates-funded GAVI vaccine alliance and Mastercard. The partnership's product is a GAVI-Mastercard Wellness Pass which is a digital vaccination record, identity system, and payment system. Mastercard has stated publicly that it is committed to promoting centralized record-keeping of childhood immunizations and professes to be the leader in a program called "World Beyond Cash." Funding for this project was matched by the Bill and Melinda Gates Foundation.[66]

The initiative uses a technology called Evergreen Hash that creates a "3D mask" based on a single picture of a person's face, palm, or fingerprint. Each person receives a "hash" that evolves over time, for example, every time a child or adult receives another vaccine.[67]

The technology is promoted as a way to help people living in areas without internet access, medical care and traditional banking; but, it is also being used for other purposes, such as "biometric voter management" in countries like Ghana. Like so many other Gates-funded projects, the recipients of the program are less enthusiastic about it than those promoting it. Many question using much-needed financial resources to re-register millions of people to vote when there are so many other more pressing issues in many of these countries, such as hunger.

Trust Stamp is also marketing its services to prisons and parole officers as a replacement for ankle bracelets, and to police departments for use in surveillance and "predictive policing." Contact tracing software has already been used to track participants in protests in response to the killing of George Floyd.[68]

Some people are already alarmed that the cashless payment systems could at some point be mandated, particularly since cash has been, for the first time, characterized as being "dirty" and a potential method for spreading COVID-19.

The ID 2020 Alliance is another initiative that launched its new digital identity program in September 2019. The Alliance is comprised of multiple partners including the government of Bangladesh, the GAVI vaccine alliance, and several other partners in government and academia. The program uses vaccination to establish digital identity and was originally promoted as a means for tracking people in third world countries who do not have birth certificates or medical records. But the city of Austin, has joined ID2020 and is working with partners to use it to "empower homeless people with their own identity data."[69]

An Israeli company founded by a former spy is now partnering with Rhode Island to use artificial intelligence to profile Americans who are already infected with COVID-19 or who are at risk of becoming infected. This information can then be given to government officials so that both individuals and the communities they live in can be targeted for mandatory testing, treatment, and lockdowns. The company says it is negotiating partnerships with several other states, hospital systems and healthcare providers, and has already signed a contract with Mayo Clinic.[70]

Is this a good idea?

There are many people who are frightened by the idea of the federal government collecting and monitoring personal health information, particularly after Edward Snowden, a former contractor for the CIA, revealed that American intelligence agencies were conducting extensive internet and phone surveillance on almost the entire population without anyone's consent.

The opposition is not to all data collection. For example, the U.S. Census is mandated by the constitution and is taken every ten years for the purpose of determining the number of seats each state will have in the U.S. House of Representatives, and also to determine where billions of dollars in federal aid will be distributed. Minimal information is collected, and it has never been used in an inappropriate manner.

There are many episodes in history, however, in which data gathering was done for more nefarious purposes. The Soviets gathered data on its citizens on several occasions and used the information to promote "re-education," collectivism and a Communist agenda.

The Nazis used the census of 1939 as a means for gathering massive data on German citizens. In partnership with Dehomag, a German subsidiary of IBM, a punch card system was used to determine, among other things, the religion of each and every German. IBM developed a Hollerith Machine to tabulate the data. A precursor to the computer, it was used to sort the cards to arrest and move Jews to concentration and death camps. The Germans used this system in other European countries as well.

At the time that these plans began, no one had any idea that the data gathered could or would be used for these purposes. Consider that the data collection and sorting technology was very primitive as compared to the technology Gates and his followers are currently proposing for Americans and their health records.

No Way Out

Perhaps the most frightening aspect of COVID-19 vaccine is the speed with which others are joining with Gates and Fauci to require that all people be vaccinated, and even that tattoos or implants be mandated.

In an article in *USA Today* on August 10, 2020 three doctors, Michael Lederman, Maxwell J. Mehlman and Stuart Youngner, state that a COVID-19 vaccine should be mandatory, and that severe penalties should be imposed on any who refuse to be vaccinated.[71] They wrote, "Private businesses could refuse to employ or serve unvaccinated individuals. Schools could refuse to allow unimmunized children to attend classes. Public and commercial transit companies — airlines, trains, and buses — could exclude refusers. Public and private auditoriums could require evidence of immunization for entry." Additionally, they wrote, tax penalties, higher insurance premiums, and even denial of both government and private services should be denied to anyone who will not get the shot; and that all religious and medical exemptions should be denied.

The doctors also endorsed the creation of a "registry of immunization," and the issuance of "certification cards" with expiration dates, which would force people to get booster shots as determined by the government.

Lederman, Mehlman and Youngner go so far as to compare the fight against COVID-19 to World War I and World War II, writing, "Everyone contributed, no one was allowed to opt out merely because it conflicted with a sense of autonomy, and draft dodgers who refused to serve were subject to penalties. True, conscientious objectors could refuse to use weapons for religious reasons, but they were obligated to help out in other ways, serving in noncombatant roles. There are no such alternatives for vaccination."

This is a ridiculous comparison. It is estimated that there were 37 million deaths attributed to World War I, and 70-85 million deaths during World War II. In mid-August 2020 there were approximately 762,000 reported COVID deaths worldwide.[72]

There is even more screening and tracking already in use...

There seems to be no end to the surveillance that is justified by COVID-19. At the Dubai airport, trained COVID-sniffing dogs are used to detect persons who are positive. The Dubai Health Authority has established a screening area in which trained canines sniff samples from armpits of travelers. If the dogs suspect COVID, passenger are then forced to take a nasal PCR test.[73]

Some states require that employers regularly conduct temperature checks, and screen for symptoms regularly. Employees can be required to undergo screening as a condition of employment.[74]

COVI-PASS is an immunity passport that can be scanned at a distance to determine a person's COVID status. It's described as a "digital health wallet" that allows "everyone the ability and peace of mind to return to work, life and travel."[75] When read it flashes green if a person has tested negative, red if the person has the virus, and yellow if it is time to be re-tested. In May, it was reported that the passes were being shipped to both governments and businesses in over fifteen countries, including Italy, France, India, and the U.S.[76]

The problems with products like COVI-PASS are horrific. We have already documented the inaccuracy of the tests, which means that hundreds of millions of people could be excluded from work, school, and general participation in society. Health information is supposed to be private, including immune status, and the passes display a person's status. Even the World Health Organization has concluded that "there is not enough evidence about the effectiveness of antibody-mediated immunity to guarantee the accuracy of an "immunity passport" or "risk-free certificate."[77]

Even schools are contemplating the use of surveillance. School officials in New Albany, Ohio announced that it was testing a system that would involve students wearing an electronic device that would track their location. The device would let

officials know if students are social distancing and allow immediate contact tracing if a student tested positive for COVID-19.[78]

In addition to the electronic tracking, the hysteria over
COVID-19 has resulted in draconian rules in the few schools that
plan to offer live classes in school buildings. Teachers and students must wear masks, children must maintain six feet distance
between desks and even when walking in hallways, there is no
recess, no lunch period, no music classes, no singing, and few or
no extracurricular activities.

There are even more rules for athletes. High school athletes
are forced to sign a "pledge" stating that they will be regularly
tested, agree to be quarantined if they are in contact with anyone
who tests positive for COVID-19 or if they themselves test positive, get a flu vaccine, wear a mask at all times, participate in
contact tracing, practice physical distancing and wash their hands
frequently as conditions of playing sports.[79]

Colleges and universities are requiring that all students
agree to random testing and will not be permitted on campus if
they refuse. At Ohio State University, faculty and staff are required to do a "health check" daily which involves answering
questions and temperature taking, after which they are issued a
check mark that allows them on campus.[80]

Enforcement!

A booklet developed by the Wisconsin Department of Health
Services called "Local Community-Isolation Site Operation
Manual" describes protocols for the operation of isolation camps

for those who are not able to self-quarantine because they live with other people. It includes job descriptions and guidelines for transporting COVID-positive patients to the facility.

These are just a few of the rules outlined in the manual:

- Individuals who are identified as needing to isolate are given 6 hours to report to the facility
- A list of what individuals are allowed to bring into the facility is provided
- Individuals are monitored with check-in calls twice daily and must answer calls
- Any packages dropped off by family and friends must be inspected by staff and objectionable items confiscated
- Meals are delivered three times per day and residents can request snacks

This sounds like instructions given to people reporting to prison.[81]

Wisconsin is not the only state that has such facilities. The National Guard in Arkansas was charged by the Governor to transport COVID-positive people to an isolation facility near Little Rock in July 2020 if they were unable to quarantine completely alone[82]

Government overreach is even worse in some other countries. In Melbourne, Australia, citizens cannot leave their homes between 8PM and 5AM. The only permissible reason for leaving one's home at any time is work, study, to shop for "essential supplies," medical care, and daily exercise for a maximum of one hour. Exercise must be performed no more than 5 km from one's home.

Only one person per household is permitted to shop, and only once per day, and shopping must be conducted within 5 km of home. All non-essential businesses are closed which is almost all businesses. No guests are permitted in homes, and weddings are not permitted. All children are learning at home; no schools are open.[83]

New Zealand experienced 100 days with no cases, and after four members of the same family were deemed probable cases, the country started locking down again. In Auckland, where the family lives, most businesses and schools are closed, and bars and restaurants can only offer takeout food. In the rest of the country, people can still go to work, and children are in school, but masks and social distancing are required.[84]

The population of Auckland is 1.6 million and the population of New Zealand is 4.8 million. There are four "probable" cases in Auckland and the city is locked down tight and the rest of the country is restricted. Is this the new normal? The world must be free of all COVID-19 cases for people to regain freedom? Freedom can be taken away as soon as there is a "probable case?"

We the people need to unite for one purpose on which we can all agree – we need to take our freedoms and liberties back again. If we do not, there will come a time when no one can protest – about anything. The only thing standing between where we are now and a complete inability to gather and protest is one more capricious and senseless order from our rulers.

ENDNOTES

1. A Partner in Shaping History – the First 50 Years. World Economic Forum https://www.weforum.org/about/history accessed 6.20.1919

2. Klaus Schwab. "Why we need the 'Davos Manifesto' for a better kind of capitalism." https://www.weforum.org/agenda/2019/12/why-we-need-the-davos-manifesto-for-better-kind-of-capitalism/ accessed 6.20.2020

3. https://www.weforum.org/partners/#B accessed 9.2.2020

4. The Great Reset. World Economic Forum. https://www.weforum.org/great-reset/ accessed 9.2.2020

5. The Great Rest: A Unique Twin Summit to Begin 2021. World Economic Forum https://www.weforum.org/great-reset/about accessed 9.2.2020

6. Klaus Schwab. Now is the time for a 'great reset'. World Economic Forum. https://www.weforum.org/agenda/2020/06/now-is-the-time-for-a-great-reset/ accessed 6.20.2020

7. Address to the Regional Committee for Europe (59th Session). World Health Organization. https://www.who.int/dg/speeches/2009/euro_regional_committee_20090815/en/ accessed 9.2.2020

8. Global Fund Partners Unite for Equitable Access to COVID-19 Tools. The Global Fund. https://www.theglobalfund.org/en/covid-19/news/2020-04-24-global-fund-and-partners-unite-for-equitable-access-to-covid-19-tools/ accessed 6.20.2020

9. Coronavirus Global Response International Pledging Event. European Commission. https://ec.europa.eu/regional_policy/en/newsroom/news/2020/05/05-04-2020-coronavirus-global-response-international-pledging-event accessed 5.23.2020

10. The EU kick-starts the Coronavirus Global Response pledging event. May 4 2020 https://www.youtube.com/watch?v=w8F2IG-7Wi68 accessed 9.2.2020

11. Maggie Fox. We need an army: Hiring of coronavirus trackers seen as key to curbing disease spread. *Stat News* April 13 2020 https://www.statnews.com/2020/04/13/coronavirus-health-agencies-need-army-of-contact-tracers/ accessed 9.2.2020

12. Andrea Mitchell Reports. The Rockefeller Foundation rolls out plan to test 30 million people a week to reopen the country. April 22 2020 https://www.msnbc.com/andrea-mitchell-reports/watch/the-rockefeller-foundation-rolls-out-plan-to-test-30-million-people-a-week-to-reopen-the-country-82402885951 accessed 9.2.2020

13. Bill Gates. What you need to know about the COVID-19 vaccine. GatesNotes April 30 2020 https://www.gatesnotes.com/Health/What-you-need-to-know-about-the-COVID-19-vaccine accessed 9.2.2020

14. Cheryl K. Chumley. Anthony Fauci sets stage for mandatory -- lucrative! – vaccine. *Washington Times* April 8 2020 https://www.washingtontimes.com/news/2020/apr/8/anthony-fauci-sets-stage-mandatory-vaccine/ accessed 9.2.2020

15. Robert Kunzia. The timetable for a coronavirus vaccine is 18 months. Experts say that's risky. *CNN* April 1 2020 https://www.cnn.com/2020/03/31/us/coronavirus-vaccine-timetable-concerns-experts-invs/index.html accessed 9.2.2020

16. Brakkton Booker. Fauci Says It's Doable To Have Millions of Doses of COVID-19 Vaccine by January. *NPR* April 3 2020 https://aequitas-inv.com/fauci-says-its-doable-to-have-millions-of-doses-of-covid-19-vaccine-by-january/ accessed 9.2.2020

17. IBID

18. Francis T, Salk J, Quilligan J. "Experience with vaccination against influenza in the spring of 1947: A Preliminary Report." *Am J Public Health Nations Health* 1947 Aug;37(8):1013-1016

19. Influenza Pandemics. The history of Vaccines. https://www.his-toryofvaccines.org/content/articles/influenza-pandemics accessed 9.2.2020

20. Jensen K, Dunn F, Robinson R. "Influenza, 1957: a variant and the pandemic." *Prog Med Virol* 1958;1:165-209

21. Langmuir A, Henderson D, Serfling R. "The epidemiologic basis for the control of influenza." *Am J Public Health Nations Health* 1964 Apr;54(4):563-571

22. Schoenbaun S, Mostow S, Dowdle W, Coleman M, Kaye H. "Studies with inactivated influenza vaccines purified by zonal centrifugation. 2. Efficacy *Bull World Health Organ* 1969;41(3-4-5):531-535

23. Gross P, Hermogenes A, Sacks H, Lau J, Levandowski R. "The efficacy of influenza vaccine in elderly persons: a meta-analysis and review of the literature." *Ann Intern Med* 1995 Oct;123(7):518-527

24. Estimates of deaths associated with seasonal influenza – United States 1976-2007. *MMWR Morb Mortal Wkly Rep* 2010 Aug;59(33):1057-1062

25. Disease Burden of Influenza. Centers for Disease Control and Prevention. http://www.cdc.gov/flu/about/disease/us_flu-related_deaths.htm accessed 9.2.2020

26. The Swine Flu Fraud of 1976 (60 Minutes with Mike Wallace). https://www.youtube.com/watch?v=Ydx_ok6gyiY&feature=youtu. be accessed 5.19.2020

27. Lasky T, Terracciano G, Magder L, et al. "The Guillain-Barré syndrome and the 1992-1993 and 1993-1994 influenza vaccines." *NEJM* 1998 Dec;339(25):1797-802.

28. Schonberger L, Bregman D, Sullivan-Bolyai J, et al. "Guillain-Barre syndrome following vaccination in the National Influenza Immunization Program, United States, 1976–1977." *Am J Epidemiol* 1979; 110(2):105–23.

29. Geier M, Geier D, Zahalsky A. "Influenza vaccination and Guillain Barre syndrome small star, filled." *Clin Immunol* 2003 May;107(2):116-21.

30. Vaccine Adverse Event Reporting System. https://vaers.hhs.gov accessed 9.2.2020

31. Adverse Effects of Vaccines: Evidence and Causality (2012). National Academies of Sciences Engineering Medicine. https://www.nap.edu/read/13164/chapter/1 accessed 9.2.2020

32. Goldman GS. Comparison of VAERS fetal-loss reports during three consecutive influenza seasons." *Hum Exp Toxicol* 2013 May;32(5):464-475

33. Simonsen L, Reichert T, Viboud C, Blackwelder W, Taylor R, Miller M. "Impact of influenza vaccination of seasonal mortality in the U.S. population." *Arch Intern Med* 2005 Feb;165(3):265-272

34. Demicheli V, Jefferson T, Al-Ansary LA, Ferroni E, Rivetti A, Di Pietrantonj C. "Vaccines for preventing influenza in healthy adults." *Cochrane Database of Systematic Reviews* 2014 Mar 13;(3):CD001269

35. Rivetti D, Jefferson T, Thomas R et al. "Vaccines for preventing influenza in the elderly." *Cochrane Database Syst Rev.* 2006 Jul 19;(3):CD004876.

36. Jefferson T, Rivetti A, Di Pietrantonj C, Demicheli V. "Vaccines for preventing influenza in healthy children." *Cochrane Database of Systematic Reviews* 2018, Issue 2. Art. No.: CD004879.

37. Shannon Brownlee and Jeanne Lenzer. Does the vaccine matter? *The Atlantic* November 2009 https://www.theatlantic.com/magazine/archive/2009/11/does-the-vaccine-matter/307723/ accessed 9.2.2020

38. Highlights of Prescribing Information FLULAVAL. http://www.fda.gov/downloads/BiologicsBloodVaccines/Vaccines/ApprovedProducts/UCM112904.pdf accessed 9.2.2020

39. Health Care-Associated Infections. U.S. Department of Health and Human Services. http://www.hhs.gov/ash/initiatives/hai/hcpflu.html accessed 9.2.2020

40. Ahmed F, Lindley M, Allred N, Weinbaum C, Grohskopf L. "Effect of Influenza Vaccination of Health Care Personnel on Morbidity and Mortality among Patients: Systematic Review and Grading of Evidence." *Clin Infect Dis.* 2014 Jan;58(1):50-57

41. Daniel O'Roark. Guest column: Influenza vaccination should never be made compulsory. *TimesNews* March 22, 2013 https://www.timesnews.net/news/local-news/guest-column-influenza-vaccination-should-never-be-made-compulsory/article_63e0cb4b-4400-5a3c-b96f-b55562b2b650.html accessed 9.2.2020

42. NJ Appeals Court Rules in Favor of Nurse who Refused Vaccine. *Health Impact News* September 3 2020 http://healthimpactnews.com/2014/nj-appeals-court-rules-in-favor-of-nurse-who-refused-flu-vaccine/

43. Cheryl Clark. Are Masks a Good Alternative to Flu Shots for Healthcare Workers? *Medpage Today* November 28 2015 http://www.medpagetoday.com/hospitalbasedmedicine/infectioncontrol/54905 accessed 9.2.2020

44. Flu and Pregnant Women. Centers for Disease Control and Prevention. https://www.cdc.gov/flu/highrisk/pregnant.htm#~:text=A%20Flu%20Vaccine%20is%20the,and%20her%20baby%20from%20flu. Accessed 9.2.2020

45. Christian LM, Iams JD, Porter K, Glaser R. "Inflammatory responses to trivalent influenza virus vaccine among pregnant women." *Vaccine*. 2011 Nov 8;29(48):8982-8987.

46. "Safety and effectiveness of FLULAVAL have not been established in pregnant women or nursing mothers." http://uprevent. mckesson.com/2855wp/wp-content/uploads/2015/09/FLULAVAL-QUADRIVALENT.pdf accessed 9.2.2020

47. Guidance for Industry. Considerations for Developmental Toxicity Studies for Preventive and Therapeutic Vaccines for Infectious Disease Indications. U.S. Department of Health and Human Services. Drug Administration. http://www.fda.gov/downloads/BiologicsBloodVaccines/GuidanceComplianceRegulatoryInformation/Guidances/Vaccines/ucm092170.pdf accessed 9.2.2020

48. Influenza Vaccination During Pregnancy. ACOG Clinical. April 2018 https://www.acog.org/clinical/clinical-guidance/committee-opinion/articles/2018/04/influenza-vaccination-during-pregnancy accessed 9.2.2020

49. Siemny Kim. Bill Gates says foundation will invest billions in fight to stop COVID-19 *KIRO7*. https://www.kiro7.com/news/local/bill-gates-says-foundation-will-invest-billions-fight-stop-covid-19/MMAFTSVGKZHPTEGYKEQKMRWTWU accessed 9.2.2020

50. Coronavirus Q&A With Anthony Fauci, MD. April 8 2020 https://www.youtube.com/watch?v=c0cYneu-hlc accessed 5.20.2020

51. Jiang S. "Don't rush to deploy COVID-19 vaccines and drugs without sufficient safety guarantees." *Nature* March 16 2020

52. Tseng CT, Sbrana E, Iwata-Yoshikawa N et al. "Immunization with SARS coronavirus vaccines leads to pulmonary immunopathology on challenge with the SARS virus." *PLoS One* 2012;7(4):e35421.

53. Yasui F, Kai C, Kitabatake M et al. "Prior immunization with severe acute respiratory syndrome (SARS)-associated coronavirus

(SARS-CoV) nucleocapsid protein causes severe pneumonia in mice infected with SARS-CoV." *J Immunol* 2008 Nov 1;181(9):6337-6348

54. Bolles M, Deming D, Long K et al. "A Double-Inactivated Severe Acute Respiratory Syndrome Coronavirus Vaccine Provides Incomplete Protection in Mice and Induces Increased Eosinophilic Proinflammatory Pulmonary Response upon Challenge." *J Vir* 2011 Dec;85(23):12201-12215

55. Robert Kunzia. "The timetable for a coronavirus vaccine is 18 months. Experts say that's risky." *CNN* April 1 2020 https://www.cnn.com/2020/03/31/us/coronavirus-vaccine-timetable-concerns-experts-invs/index.html accessed 9.2.2020

56. Moderna, NIAID Partner and Planned Trial of Coronavirus mRNA Vaccine. *Genetic Engineering and Biotechnology News* Feb 25 2020 https://www.genengnews.com/news/moderna-niaid-partner-on-planned-trial-of-coronavirus-mrna-vaccine/ accessed 9.2.2020

57. Prefusion coronavirus spike proteins and their use. US20200061185A1 https://patents.google.com/patent/US20200061185A1/en accessed 5.20.2020

58. https://www.sec.gov/Archives/edgar/data/1682852/000119312520001732/d117801dex991.htm accessed 5.20.2020

59. Moderna, NIAID Partner and Planned Trial of Coronavirus mRNA Vaccine." *Genetic Engineering and Biotechnology News* Feb 25 2020 https://www.genengnews.com/news/moderna-niaid-partner-on-planned-trial-of-coronavirus-mrna-vaccine/ accessed 9.2.2020

60. Eric Boodman. "Researchers rush to test coronavirus vaccine in people without knowing how well it works in animals." *STAT* March 11 2020 https://www.statnews.com/2020/03/11/

researchers-rush-to-start-moderna-coronavirus-vaccine-trial-with-out-usual-animal-testing/ accessed 9.2.2020

61. Bill Gates. What you need to know about the COVID-19 Vaccine. Humankind has never had a more urgent task than creating broad immunity for coronavirus. *GatesNotes* April 30 2020 https://www.gatesnotes.com/Health/What-you-need-to-know-about-the-COVID-19-vaccine?WT.mc_id=20200430165003_COVID-19-vaccine_BG-TW&WT.tsrc=BGTW&linkId=87665522 accessed 9.2.2020

62. Mike Williams. "Quantum-dot tattoos hold vaccination record." Rice University News and Media Relations. December 18 2019 http://news.rice.edu/2019/12/18/quantum-dot-tattoos-hold-vaccination-record/#:~:text=When%20the%20needles%20dissolve%20in,biocompatible%2C%20micron%2Dscale%20capsules. Accessed 9.2.2020

63. Kim E, Erdos G, Huang S et al. "Microneedle array delivered recombinant coronavirus vaccines: Immunogenicity and rapid translational development." *EBioMedicine* 2020 May;55:102743

64. McHugh KJ, Jing L, Severt SY et al. "Biocompatible near-infrared quantum dots delivered to the skin by microneedle patches record vaccination" *Sci Trans Med* 2019 Dec;11(523):eaay7162

65. https://www.cnn.com/videos/health/2020/04/10/dr-anthony-fauci-coronavirus-full-interview-newday-vpx.cnn accessed 9.2.2020

66. Raul Diego. "Africa to Become Testing Ground for "Trust Stamp" Vaccine Record and Payment System." *BigTechtopia* July 16 2020 http://bigtechtopia.com/2020/07/africa-to-become-testing-ground-for-vaccine-record-and-payment-system/ accessed 9.2.2020

67. Kristin Klobersanz. "Signed, Sealed, encrypted: This digital ID is all yours." *Startup Stories* June 26 2020 https://mastercardcontentexchange.com/perspectives/2020/signed-sealed-encrypted-this-digital-id-is-all-yours/ https://mastercardcontentexchange.com/

perspectives/2020/signed-sealed-encrypted-this-digital-id-is-all-yours/ accessed 9.2.2020

68. Andy Meek. "Minnesota is now using contact tracing to track protesters as demonstrations escalate." *BGR* May 30 2020 https://bgr.com/2020/05/30/minnesota-protest-contact-tracing-used-to-track-demonstrators/ accessed 9.2.2020

69. Chris Burt. "ID2020 and partners launch program to provide digital ID with vaccines." *Biometricl Update.com* Sept 20 2019 https://www.biometricupdate.com/201909/id2020-and-partners-launch-program-to-provide-digital-id-with-vaccines accessed 5.23.2020

70. Whitney Webb. "Meet the Israeli Intelligence-Linked Firm Using AI to Profile Americans and Guide US Lockdown Policy." July 2 2020 https://www.thelastamericanvagabond.com/meet-israeli-intelligence-linked-firm-using-ai-profile-americans-guide-us-lockdown-policy/

71. Dr. Michael Lederman, Maxwell J. Mehlman, Stuart Youngner. "Defeat CVID-19 by requiring vaccination for all. It's not un-American, it's patriotic." *USA Today* August 10 2020 https://www.usatoday.com/story/opinion/2020/08/06/stop-coronavirus-compulsory-universal-vaccination-column/3289948001/ accessed 9.2.2020

72. https://ourworldindata.org/covid-deaths accessed 8.10.2020

73. Janine Puhak. Coronavirus-sniffing dogs deployed at Dubai Airport with 91% accuracy: report. *Fox News* August 14 2020 https://www.foxnews.com/travel/coronavirus-sniffing-dogs-dubai-airport accessed 9.2.2020

74. Temperature Screening: New Guidance From the CDC, FAQs, and Best Practices. The National Law Review. May 1 2020 https://www.natlawreview.com/article/temperature-screening-new-guidance-cdc-faqs-and-best-practices accessed 8.14.2020

75. https://covipass.com/ accessed 8.10.2020

76. Susan Halpern. Immunity Passports and the Perils of Conferring Coronavirus Status. *The New Yorker* May 22 2020 https://www.newyorker.com/tech/annals-of-technology/immunity-passports-and-the-perils-of-conferring-coronavirus-status accessed 9.2.200

77. World Health Organization. "Immunity passports" in the context of COVID-19. https://www.who.int/news-room/commentaries/detail/immunity-passports-in-the-context-of-covid-19 accessed 8.8.2020

78. Schools Turn to Surveillance Tech to Prevent COVID-19 Spread. *Wired* June 5 2020 https://www.wired.com/story/schools-surveillance-tech-prevent-covid-19-spread/ accessed 9.2.2020

79. Ohio High School Athletic Association. Ohio High School Athletic Association Acknowledgement and Pledge. www.ohsaa.org

80. Brittany Bailey. COVID-19 testing now required for all Ohio State students in campus housing. *WBNS News* August 12 2020 https://www.10tv.com/article/news/local/covid-19-testing-now-required-for-all-osu-students-in-campus-housing/530-29d2b4dd-1000-4f59-8699-fc02f9c7b4c3 accessed 9.2.2020

81. COVID-19. Local Community – Isolation Site Operation Manual P-02639 (6/11/2020)

82. Arkansas National Guard transporting COVID-19 patients to isolation facility. *Fox13* July 13 2020 https://www.fox13memphis.com/news/local/arkansas-national-guard-transporting-covid-19-patients-isolation-facility/M6QOEDIC4JEGLKO7L2AN4BVYDQ/ accessed 9.2.2020

83. COVID-19: Victoria's new coronavirus rules, explained. https://www.racv.com.au/royalauto/living/community/victoria-new-coronavirus-rules-explained.html accessed 9.2.2020

84. Lauren Wamsley. "New Outbreak in New Zealand Leads to New Rules and Supermarket Runs." https://www.npr.org/sections/coronavirus-live-updates/2020/08/12/901745392/new-outbreak-in-new-zealand-leads-to-new-rules-and-supermarket-runs accessed 9.2.2020

REFLECTIONS ON THE COVID-19 DEBACLE

WE HAVE SOME GOOD news and some bad news. Let's start with the bad.

Our government has collapsed and has been taken over by illegitimate actors. They are despots who at one time were elected officials, and incompetent imposters who describe themselves as health experts. These people exercise control over their subjects by issuing ridiculous rules and regulations that would have been considered unconstitutional under our old system of government.

Our medical system has collapsed. Hospitals have lost hundreds of billions of dollars, doctors and nurses have been laid off and services are limited. Rules that include incessant COVID-19 testing and complete disconnection from family and friends in order to receive care are a disincentive for almost anyone who is not on the verge of death to seek medical treatment.

Our education system has collapsed. Most children stopped learning anything by the end of March. Teachers unions claim that their members do not want to teach. Rules for the opening and operation of schools have resulted in an environment that is psychologically unsafe for children and adolescents.

The media has destroyed itself and lost all legitimacy. Few journalists have asked questions and tried to expose the COVID hoax. Most have dutifully reported anything the illegitimate government tells them to report, regardless of the clearly preposterous nature of the information. The media are now puppets of the regime.

The population is divided between thinking people and "the others," those who wear the mask with a glazed look in their eyes and believe what the regime and its puppet media report to them every day. These factions can no longer engage in conversation. The line in the sand has been drawn. You believe or you do not believe what the government reports daily about COVID-19.

Were Things Really Better Before?

Many people think back to February of this year and long for a return to that time, when things were "normal." We used to. We don't any longer.

There were some things that were nice in February. We lived in a free society. We did what we wanted to when we wanted to do it (of course within reasonable legal limits). We had access to libraries and museums, plays and symphony concerts, dance and other culture. We could travel to almost anywhere in the world and enjoy new experiences.

But something else was going on in February. A perfect storm was brewing. Government had gradually grabbed more and more power. Citizens had become lazy and allowed the government to intrude in their private lives. The media had been corrupted and regularly reported false information. The education system had deteriorated to the place where illiterates could graduate from high school and colleges protected students from information that contradicted their beliefs. Perhaps worst of all, a cartel had formed comprised of Big Medicine and Big Pharma, and academic institutions and government agencies funded by industry. Research shows that hundreds of thousands of Americans were being harmed and killed by this cartel every year, and no one thought that this was going to change, at least not during our lifetimes.

Finally, most people, including politicians, had lost any ability to discuss political differences or work toward compromise. Politics had become a game where the winner takes all, and the end justifies any means.

So, we don't long for the life we had in February. It was unsustainable, and sooner or later it was going to blow up. It did blow up and here we are. We are being held prisoner by a hostile regime. We have lost many rights. Our society has collapsed. We have fallen into the abyss. We cannot go back to the way it was before.

What Do We Do Now?

There is a silver lining in this very thick, dense cloud. We can recover. Not by trying to go back to where we were, but rather by creating new systems based on our values and what we, the people, want. It is sometimes easier to build something new than to fix something old and dysfunctional.

Let's start with education. Politicians and educators have been talking about reform for decades. New plans and "improvements" were adopted every few years, all of which resulted in the system getting worse. And there was no end in sight.

What is happening now? Almost 40% of parents are not sending their children to school this year. Most are not afraid of "the virus." Rather they are looking to avoid the dystopian nightmare that awaits their kids at school, and to provide better education. Pods and co-ops are forming. New online schools are opening, and parents are choosing the educational pathways they think is best. This is real school choice, and real reform. We are actively promoting this and working hard to help parents who want to home school. School systems will either change and respond to demands from parents, or more and more parents will take education into their own hands and home school. The old system is gone and never coming back. This is good and would never have happened without the fake pandemic.

How about government? Criminals and despots have taken over our government, and this is a good thing. These were bad people before, and we just did not see it so clearly. Now we do and we think most thinking people want these folks out of office. Recall petitions and elections are the way to do this. People are asking politicians the right questions and making choices based on their answers. Members of both parties are crossing over to vote for candidates who represent their values and meet their expectations. More good people will end up in office with a desire to serve the public instead of self-aggrandizement. This is so much better!

What about our healthcare system? One of the biggest impediments to medical reform has been the consolidation of the medical profession, with an increasing number of healthcare providers working in large institutions. These institutions have dictated how medicine was practiced and stifled innovation. There are now tens of thousands of unemployed doctors and nurses, some of whom may decide to start their own practices. We would like to see the return of the independent family practice doctor like the ones who took care of our families when we were kids.

Another even more important outcome can result from this mess. Trust in doctors and medicine is one of the reasons why people have been harmed by medical care. For decades, people have been brainwashed into thinking that doctors know best, that medicine is too complicated for mere mortals to understand, and that following a doctor's instructions leads to better health. The continuation of the medical cartel as described above is dependent on this belief.

Fortunately, millions of people have awakened due to the fake pandemic. They have watched the country's "best doctors" – the ones who have made the important decisions about this debacle – demonstrate that they are at least incompetent and quite possibly criminally liable. They have seen public health officials in most states misrepresent data and issue orders that have hurt more people than could possibly be helped. They have listened to hundreds of doctors who appeared on news programs dutifully repeating lies told to them by the CDC and WHO. Untold numbers of health professionals have participated in the fake testing and falsification of death certificates. Belief and trust in medicine have been shattered, new standards are needed, and consumers are more aware of the importance of being informed.

Creating Our New Future

This is the time for all people from all races and ethnicities, from all religions, and from all political parties to come together to build a new world. We are not talking about the "new normal" that the criminals and despots have in mind for us. Rather the one we want. Government accountable to citizens. Outcomes-based education for our children. A medical system that respects consumer choice and serves us. We can have all of this and more if we focus on the things we have in common. A common desire to be a free people and set aside our differences in order to achieve this very important goal.

Let's do this. We can do this. We want to do this, and we hope you will join us.

www.makeamericansfreeagain.com

OUR NEW NORMAL: CITIZENS IN CHARGE

All citizens of the world must wake up and act now!

MANY WELL-MEANING people are circulating petitions and sending emails to government leaders. Detailed and well-researched information has been presented to our rulers as evidence that their policies are ill advised and harmful. As we have seen, the despots are not interested in science, and in fact are not interested in our welfare at all.

We are not going to change the minds of people who can hurt us in such egregious ways.

There is something we can do. We can unite! The tens of millions of Americans who are fed up with this insanity can gather and declare our independence from tyranny. **Our American initiative is called MAKE AMERICANS FREE AGAIN** and we hope that it will become the template for other countries to declare freedom from their rulers as well.

With this initiative, we the people are announcing a new normal to the despots – it is called Citizens in Charge. Politicians are not the boss of us. They only have jobs because we elect them, and we can un-elect them. Health officials have no authority over our lives either. **Here is a message for all the despots – and you know who you are – your 15 minutes are up. Citizens are now in charge!**

Acting now flips the script on the despots and sends the clear message we are done with their precedent-setting. We the people must set the precedent of standing up for our freedoms. While the following points are specific to the current situation, it is entirely appropriate to consider these actions at any point in the future when government officials forget their role – which is to represent and do the work for the people!

To accomplish our goal, we are forming a huge group of Americans around a simple, easily understandable platform:

"I want the freedom to accept or reject vaccination for myself and my minor children. The state or federal government cannot force me or my minor children to be vaccinated without my express permission."

We are one-issue voters. We agree to vote for politicians who are for our platform and we vote out those who are not. We make this choice regardless of party affiliation and without any consideration for other political issues, understanding that there is no issue more important than the basic freedom from unwanted and potentially dangerous medical intervention.

If you agree with this platform, you join our group. If not, you don't. Simplicity is important.

As we develop our database of people who are of like mind, we will help them connect with one another at the local level to fight against remaining COVID-related restrictions. We have seen that large numbers of people defying lockdown restrictions usually get away with it. A great example is the recent protests and riots. Thousands of people were in one place without social distancing and nothing happened.

Others include entire counties that have notified their rulers that they are opening up, sheriffs who have informed the despots that they will not enforce the rules, and huge churches that have held services in packed rooms with singing allowed.

There is strength in numbers, and people need to be able to find like-minded people so that they can join forces, pressure government officials, and even engage in civil disobedience if necessary.

The despots have not yet built concentration camps, usually referred to as re-education centers in other countries ruled by despots. This is most likely coming soon, however, so we need to act now!

It is especially important that we stay razor focused on the simplicity of our approach. Everyone can understand it. Everyone can explain it. It is a yes or no question for a politician. Simplicity is key since we have already seen that the complicated approach does not work.

We have also planned activities that we think are important to the types of people who are interested in our simple

platform – freedom from unwanted and harmful medical intervention.

We will provide resources and guidance for parents who decide to withdraw their children from schools (different than schooling at home as directed by despots while your children are still enrolled in public or private schools).

Studies show that 40% of parents are considering this, and more would do it if they knew how. There are a couple of reasons for this. Many parents have discovered that their kids were not learning much at school, and the psychological damage that will result from the draconian plans the schools have developed will last for a lifetime. These include the requirement to wear masks, stay 6 feet apart from other children, no shared supplies, no lunch, no recess, and riding to school on a bus with one child in every other row.

Some states have already passed laws mandating that children receive dozens of vaccines as a prerequisite for enrollment. Many are considering adding mandatory COVID-19 vaccines when they are available, too.

Parents need assistance in choosing online schools and/or curricula; identifying parents, adults, or babysitters who can supervise children whose parents work outside the home; and arranging local activities for home-schooled children like social gatherings, sports, arts and theater classes, and field trips. Home schooled kids do not lack social interaction, or any other benefits of traditional schools. We have built a resource center to assist parents in making this transition.

In addition to providing better education and protecting children from ridiculous rules and vaccine mandates, increased home schooling will provide another great benefit. School districts receive funding, in part, based on enrollment. Declining enrollment means less funding, and some buildings may even have to close. While many parents may choose to continue to home school regardless of what schools do, some systems may figure out that they cannot ignore what local parents want and become much more accommodating.

We are developing a network of lawyers who will represent people whose civil rights are being violated by any government official or agency, and who will organize class action lawsuits when warranted. Lawsuits for loss of income, destruction of businesses, wrongful death, mental and emotional abuse, and violation of civil rights are just a few of the lawsuits currently being filed. Families of children who have gone backwards academically should sue. Until we can get rid of the despots, we need to keep them busy in court day after day after day and find ways to go after them personally. Government officials often do bad things and they rarely face consequences. This time they have stepped so far over the line that many are most likely personally liable for what they have done.

Once we regain our freedom, we collectively agree to help businesses that are struggling to survive. We purchase anything we can from them, offer business advice, financial support when possible, and assistance in relocating, negotiating with landlords and other help as needed.

Rationale for Doing Things This Way

Why do things this way instead of trying to persuade politicians based on science? Enormous amounts of work and money have been invested in trying to prevent and reverse vaccine mandates. Well-researched documents have been prepared for legislators; health professionals and other experts have provided exceptional testimony; and parents have told horrifying stories about their injured children. The problem is that for every document the pro-freedom side presents, the other side presents one too. For every health professional or researcher who provides testimony, mandate advocates present one to say the opposite. For every consumer who offers a story in support of freedom of choice, a consumer who insists that "anti-vaxxers" are a danger to her children tells a story too. At the end of the day there is no winner. Petitions and emails also have little impact.

A major barrier is the gargantuan funding drug companies and medical associations give to political campaigns and political action committees (PACs). It is reasonable to assume that these entities expect a return on their investment. Historically their investment seems to have paid off; thus, they continue to support politicians who support their agenda.

The limitation of political contributions is that they cannot buy votes. And what politicians want more than anything is to remain in office. Thus, blocks of voters, if large enough, can be formidable.

Another problem is the complexity of the vaccine issue, with competing experts, confusing research, and the ability of elected officials to claim that more research is needed, more time is required for thought, or more people need to be consulted before decisions can be made.

Thus, the key to swaying politicians is to make the proposition simple: "I want the freedom to accept or reject vaccination for myself and my minor children. The state or federal government cannot force me or my minor children to be vaccinated without my express permission." Are you a yes or a no on this issue? If you choose not to answer, our group will assume the answer is no."

It is common for state representatives and senators to win their races by only a few hundred or a few thousand votes, which means that a group of us in a district equal to the winning margin in the last election can be very threatening to a politician's future. It is not necessary to target all pro-vaccine politicians in a legislative body. Our success in a few districts can be expected to influence others who do not wish to lose their elections over this issue.

We have a limited time to take our country back!
It is time to act now!

EPILOGUE

THERE ARE MANY unanswered questions about COVID-19. Was the virus man-made? Did the Chinese deliberately release the virus to create this crisis? Is COVID-19 actually a virus? Did Bill Gates and Anthony Fauci collaborate to spread false information and promote fear? Were the governors aware of the hoax and willingly play along? If so, why?

All these issues are worth investigating, and perhaps we will write another book when we know the answers to these questions. What we know now is that this "pandemic" was used for the purpose of social engineering. We should not be surprised. While we humans pride ourselves in our ability to engage in critical thinking and to make deliberate choices, in recent years, we have allowed government to become more and more powerful and to take more and more of our freedoms away. COVID-19 is just the biggest power grab in history, affecting the world's entire population.

Throughout this grand experiment, we have essentially all been lab rats for the rich and powerful who tell us that they know what is best for us. These people have directed our rulers to experiment with how much control and how many rules we will tolerate. We are told regularly that we must follow directions and that we will be punished if we do not. We follow directions and the restrictions are loosened a bit with warnings about the need to be compliant. Invariably, it is determined that 100% of the population is not compliant and they punish us by taking away some of our privileges again.

Those conducting this experiment are most likely happy with the results. Most people have complied. Already many people are so grateful to be able to see friends or buy shoes, or take their children to the park that they forget that the right to do these things is granted in our constitution, not by the despots who have taken over our government and our lives.

This experiment can continue only for as long as we allow it. We the people must demonstrate that we will no longer participate. Protests and lawsuits are important. The most important step we can take is to just stop this. ALL of this!

Stop being afraid. Take off the mask. Open your business. Walk the wrong way down the aisles of the grocery store. Have a party – a big one. Find a large landowner who will allow you to schedule weekend concerts for thousands of people – every weekend. Start hugging people – in public. Start every day with the intention of breaking the rules – more and more of them.

The rulers cannot do anything to stop millions of people who decide to do this. The experiment will be over. One more time, humanity will have demonstrated that the dark forces of evil cannot prevail.

We must not relax, because if we do, this will most certainly happen again. A few nefarious people decided to invent a public health emergency and managed to control the world for several months. This was almost certainly a "practice run" for more permanent rule. It is up to us to make sure that does not take place.

So, our work has only just begun. The power-hungry politicians who manage to remain in office must be forced to resign or voted out. Some, along with the wealthy individuals and organizations that directed them, should be criminally indicted for crimes against humanity. Laws must be passed to set guidelines for declaring an emergency and to establish limitations on how long such a declaration can last without legislative oversight.

Perhaps most important, we need to pause and reflect on the importance of freedom, and our individual and collective responsibility for maintaining it. Americans should invest time in learning about our history, our constitution, and our rights and responsibilities as citizens. We must teach these principles to our children. We must never again take our liberty for granted.

CPSIA information can be obtained
at www.ICGtesting.com
Printed in the USA
FSHW010515211020
75075FS